Praise for *BOLD*

"*BOLD* is a visionary roadmap for people who believe they can change the world—and offers invaluable advice about bringing together the partners and technologies to help them do it."

—President Bill Clinton

"A guide to exponential digital chutzpah from a master of the art of 'going big.'"

—*Financial Times*

"*BOLD* is an essential navigation tool for any proactive CEO who wants to remain relevant. In the next decade it is reasonable to assume that some of the corporations at the top of the Fortune 500 will be displaced by the 'exponential entrepreneur.' History tells us that if we don't proactively change and adapt, change will be imposed on us. *BOLD*, spells out the implications and opportunities driven by exponential changes transforming our world."

—Jim Moffatt, CEO, U.S. Deloitte Consulting, LLP

"Expressed with sunny optimism and promise, Diamandis and Kotler share their extensive experience and knowledge, hoping to boost innovative potential within the technology startup arena and inspire readers to 'get off the couch and change the world.' An empowering and multifaceted 'playbook' for the creative entrepreneur."

—Kirkus Reviews

"*Abundance* showed us where our world can be in 20 years. *BOLD* is a roadmap for entrepreneurs to help us get there."

—Eric Schmidt, executive chairman, Google

"This is a manual for today's big thinkers to become tomorrow's bold leaders, using crowd-powered tools accessible to everyone."

—*Booklist*

"If you read one business book in the twenty-first century, this should be the one. When Peter and I cofounded Singularity University, we based it on the ideas of exponential change and 'learn by doing.' This book clearly explains how to apply these concepts to change the world and overcome the age old afflictions of human civilization."

—Ray Kurzweil, inventor, author, director of engineering at Google, chancellor of Singularity University

"This invigorating discussion drives home the point that with better tools than we've ever had before, what we need most of all are great leaders."

—*Publishers Weekly*

"In *BOLD*, Diamandis and Kotler have written another dazzler. A riveting look inside the world of exponential entrepreneurship—action-packed and action-oriented. I've purchased a copy for my entire team at Cisco."

—Padmasree Warrior, CTO & chief strategy officer, Cisco

"It makes bold predictions and teaches entrepreneurs how to thrive in the same way as our mammalian ancestors: by being nimble and resilient."

—HuffingtonPost.com

"I loved Peter Diamandis and Steven Kotler's book *Abundance*, their writing and their vision. *BOLD* is an amazing sequel, a book that every entrepreneur should read. It is inspiring, filled with incredible insights and offers a practical how-to game plan for going big and impacting the world."

—Michael Dell, CEO, Dell Computers

"The infectious optimism of BOLD is inspirational."

—*New York Times* "DealBook" Blog

BOLD

HOW TO GO BIG, ACHIEVE SUCCESS, AND IMPACT THE WORLD

PETER H. DIAMANDIS AND STEVEN KOTLER

SIMON & SCHUSTER PAPERBACKS

New York   London   Toronto   Sydney   New Delhi

Simon & Schuster Paperbacks
An Imprint of Simon & Schuster, Inc.
1230 Avenue of the Americas
New York, NY 10020

First Simon & Schuster trade paperback edition February 2016

SIMON & SCHUSTER PAPERBACKS and colophon are registered
trademarks of Simon & Schuster, Inc.

For information about special discounts for bulk purchases,
please contact Simon & Schuster Special Sales at 1-866-506-1949
or business@simonandschuster.com.

The Simon & Schuster Speakers Bureau can bring authors to your live event.
For more information or to book an event, contact the Simon & Schuster Speakers
Bureau at 1-866-248-3049 or visit our website at www.simonspeakers.com.

Manufactured in the United States of America

1   3   5   7   9   10   8   6   4   2

Library of Congress Cataloging-in-Publication Data is available.

ISBN 978-1-4767-0956-7
ISBN 978-1-4767-0958-1 (pbk)
ISBN 978-1-4767-0960-4 (ebook)

CONTENTS

PETER'S DEDICATION

I dedicate this book to my parents,
Harry P. Diamandis, MD and Tula Diamandis,
whose bold journey from the Greek island of Lesvos to the
United States, and their success in medicine and family
inspired me to go big, create wealth, and impact the world.

STEVEN'S DEDICATION

This one is for Jamie Wheal,
my great friend and partner in the Flow Genome Project,
without whom this journey would be a lot less
interesting and make a lot less sense.

INTRODUCTION

Birth of the Exponential Entrepreneur

Go back some 66 million years, and life on Earth was a little different. These were the waning days of the Cretaceous Period, hot and humid, when much of the world's current land mass was still submerged under massive oceans. Back then, angiosperms, our technical name for flowering plants, were the latest innovation in the world of flora. Similarly, our first maples, oaks, and beeches were just starting to emerge. On the fauna side, the Earth was still dinosaur dominated, but this is not surprising. When it comes to staying power, the 100 million years that these megareptiles lorded over our planet's terra firma is the longest such stretch in history—the ultimate example of terrestrial dominance.[1]

But their reign was not to last.

The Cretaceous Period ended with a very big bang.[2] An asteroid some ten kilometers in diameter—or slightly smaller than San Francisco—smashed into the Yucatán Peninsula in Mexico. The collision literally rocked the world, releasing 420 zettajoules of energy, or two million times more muscle than the largest nuclear bomb ever exploded. The resulting crater was 110 miles wide. The resulting impact was, as the saying goes, "a planetary killer."

Megatsunamis, massive earthquakes, global firestorms, and a

deadly cascade of volcanic eruptions swallowed the Earth. The sun disappeared behind a huge dust cloud—and didn't emerge for a decade. The changes to the global environment were so rapid and so extreme that the dinosaurs—the uberdominant form of life at the time—were unable to adapt. Instead, they went extinct.

For our species, this was very good news. While the dinosaurs were large, lumbering, and inflexible, those early small, furry mammals—our ancestors—were far more nimble and resilient. They took opportunistic advantage of the radical changes sweeping the globe, adapted to their new environment, and never looked back. Within an evolutionary eye blink, the dinosaurs were gone and mammals became kings of the world. And one thing is most certain—history has a funny way of repeating itself.

In fact, this tale of colossal impact, radical transformation, and spectacular rebirth has exceptional relevance today—especially for business. Right now, there is another asteroid striking our world, already extinguishing the large and lumbering, already clearing a giant path for the quick and nimble. Our name for this asteroid is "exponential technology," and even if this name is unfamiliar, its impact is not.

We'll get into far greater detail later, but what's important here is that exponential technology refers to any technology accelerating on an exponential growth curve—that is, doubling in power on a regular basis (semiannually, annually, etc.)—with computing being the most familiar example. When a woman in Outer Mongolia answers her smartphone, she's using a device a million times cheaper and a thousand times more powerful than a supercomputer from the 1970s.[3] That's what exponential change looks like in the real world.

And today, this kind of change is everywhere we look. Exponential progress is now showing up in dozens of arenas: networks, sensors, robotics, artificial intelligence, synthetic biology, genomics, digital medicine, nanotechnology—to name only a few.[4] And like our ten-kilometer asteroid, the awesome power of these technologies is reshaping life on Earth. But this same power is threatening a different breed of dinosaur—those large and innovation-resistant companies that have done it the same way for decades, and will continue to do it the same way, until, well, they are out of business.

Yet, in stark contrast, there is a new breed of small, furry mammal starting to emerge. These mammals are today's entrepreneurs— the ones using radically accelerating technology to transform products, services, and industries. These nimble and resilient innovators are learning how to wield exponential technologies; they are becoming *exponential entrepreneurs*. And these exponential entrepreneurs are paving the way for a new world of abundance.

The Follow-on to Abundance

In 2012, I joined with Steven Kotler to write *Abundance: The Future Is Better Than You Think*. My inspiration for this book came from my work with both the XPRIZE Foundation and Singularity University. What I witnessed from the helm of those organizations was a world where the basic necessities of life were becoming cheaper and globally accessible. Steven brought to *Abundance* his considerable expertise mapping the intersection of ultimate human performance and exponential technology. Both of us had come to believe the world was radically changing and that, for the first time in history, humanity had the potential to significantly and permanently raise global standards of living.

In *Abundance*, Steven and I explored how four potent emerging forces—exponential technologies, the DIY innovator, technophilanthropists, and the rising billion—give us the ability to solve many of the world's grandest challenges over the next two to three decades. That is, we will soon have the power to meet and exceed the basic needs of every man, woman, and child on the planet.

When the book was released in February 2012, we had little idea how it would be received. I was lucky enough to open the TED conference with a talk on *Abundance*—and even luckier to get a standing ovation. The book rocketed onto the charts, spent almost three months on the *New York Times* bestseller list, won several "Best Book of 2012" awards,[5] and has been translated into more than twenty languages. For all of this, we are both incredibly grateful.

What's also been deeply gratifying is that concrete, well-documented evidence for abundance continues to mount. As a result, in the 2014 paperback edition of *Abundance*, we proudly presented a new reference section that contains some sixty additional charts, in areas such as reduction of violence and increases in learning, health, and wealth. Taken together, the implications of this data are truly mind-blowing.

But we have also come to feel that painting a picture of our vibrant future is insufficient. While we truly believe that creating a world of abundance is possible, it is by no means guaranteed. And it is for that reason we wrote *Bold*.

The World's Biggest Problems = Biggest Business Opportunities

Thousands of years ago, it was only kings, pharaohs, and emperors who had the ability to solve large-scale problems. Hundreds of years ago, this power expanded to the industrialists who built our transportation systems and financial institutions. But today, the ability to solve such problems has been thoroughly democratized. Right now, and for the first time ever, a passionate and committed individual has access to the technology, minds, and capital required to take on any challenge. Even better, that individual has good reason to take on such challenges. As we will soon see, the world's biggest problems are now the world's biggest business opportunities. This means, for exponential entrepreneurs, finding a significant challenge is a meaningful road to wealth. Ultimately, as I teach at Singularity University (much more on this later), the best way to become a billionaire is to solve a billion-person problem.

In *Bold*, Steven and I offer a highly practical playbook for doing just that. This book arms today's entrepreneurs, activists, and leaders with the tools needed to positively impact the world while simultaneously making their biggest dreams come true. To make good on that promise, *Bold* unfolds in three parts: Part One focuses on the exponential technologies which are disrupting today's Fortune 500 and enabling upstart entrepreneurs to go from "I've got a new idea" to "I run a billion-dollar company" far faster than ever before. Part Two of the book focuses on the psychology of Bold—the mental tool kit that allows the world's top innovators to raise their game by thinking at scale—and includes detailed advice and lessons from technology gurus such as Larry Page, Elon Musk, Richard Branson, and Jeff Bezos. Also in Part Two, Steven reveals the keys to ultimate human performance garnered from fifteen years of research with the Flow Genome Project, and I reveal my entrepreneurial secrets garnered from starting seventeen companies. Finally, Part Three closes the book with a look at the incredible power and essential best practices that allow anyone to leverage today's hyperconnected crowd like never before. Here you'll learn how to harness crowdsourcing solutions to massively increase the speed of your business, to design and use incentive competitions to find breakthrough solutions, to launch million-dollar crowdfunding campaigns to tap into tens of billions of dollars of available capital, and finally, to build exponential communities—armies of exponentially enabled individuals willing and able to help today's entrepreneurs make their boldest dreams come true.

Who Should Read This Book?

This book was written as both a manifesto and a manual for today's exponential entrepreneur, anyone interested in going big, creating wealth, and impacting the world. It is a go-to resource on accelerating technologies, thinking at scale, and utilizing crowd-powered tools. If you are an entrepreneur, in spirit or by experience, whether you live in

Silicon Valley or Shanghai, whether you are in college or an employee
of a multinational corporation, this book is for you. It's about seriously leveling up your abilities and your ambition. It is about moonshot thinking and global impact.

If, on the other hand, you are a manager, executive, or owner of the
large and lumbering, your competition is no longer some multinational
from overseas—it's now the explosion of exponential entrepreneurs
working out of their garages. Reading this book will give you insight
into where this new competition is coming from and how they think
and operate. Moreover, the same exponential opportunities—meaning
both the technologies themselves and the strategies for maximizing
these technologies (both psychological and organizational)—exist for
solo entrepreneurs and big companies. Finally, if you're an organizational leader and are interested in going even deeper into this subject,
then I recommend reading Singularity University's first publication:
Exponential Organizations (ExO), written by Salim Ismail, SU's first
executive director and current global ambassador. *ExO* is written for
the leadership of those companies that prefer to sidestep extinction
and join the exponential revolution.

Last, and perhaps most importantly, *Bold* is a playbook. Our deep-
est hope is that it inspires you to get off the couch and change the
world. Said differently, because of the amazing opportunities created by exponentially growing communications technology, many of
today's best and brightest have been lured in by an app-tilted playing field, which has both entrepreneurs and venture capitalists believ-
ing that three years to profitability and exit should be the norm. Of
course, if your true passion is building apps, then build away. But
let's be clear: when Steve Jobs said that the goal of every entrepreneur
should be to "put a dent in the universe"—he wasn't talking about
inventing the next Angry Birds. This book is for those who want to
make the Giant Dent. It's about the fact that, because of exponential
empowerment, anyone can make that Giant Dent. Seriously, what are
you waiting for?

A Collaboration of Two Minds

Peter and Steven first met in 1997, when Steven wrote a feature about the XPRIZE. In the late 2000s, they came together to write *Abundance: The Future Is Better Than You Think*. Upon its success, Peter approached Steven with the concept for *Bold*, and asked him to team up once again and write a book focused on inspiring and enabling entrepreneurs to create this world of abundance. Once again, both Peter and Steven brought their unique perspective and expertise to the table. So, while this book is told in Peter's voice and through his stories, this work is a true partnership, as the ideas and the writing in *Bold* were shared equally between Peter and Steven.

—Peter H. Diamandis
Santa Monica, California

—Steven Kotler
Chimayo, New Mexico

PART ONE

BOLD TECHNOLOGY

CHAPTER ONE

Good-bye, Linear Thinking . . . Hello, Exponential

Birth of a Behemoth

The year was 1878. George Eastman was a twenty-four-year-old junior clerk at the Rochester Savings Bank in need of a vacation. He chose to go to Santo Domingo, in the Dominican Republic. At the suggestion of a coworker, Eastman bought all the requisite photographic equipment to make a record of the trip. It was a lot of equipment: a camera as big as a Rottweiler, a massive tripod, a jug of water, a heavy plateholder, the plates themselves, glass tanks, an assortment of chemicals, and, of course, a large tent—this last item providing a dark place in which to spread emulsion on the plates before exposure and a dark place to develop them afterwards. Eastman never did go on that vacation.[1]

Instead, he got obsessed with chemistry. Back then photography was a "wet" art, but Eastman, who craved a more portable process, read about gelatin emulsions capable of remaining light-sensitive after drying. Working at night, in his mother's kitchen, he began to experiment with his own varieties. A natural-born tinkerer, Eastman took less than two years to invent both a dry plate formula and a machine that fabricated dry plates. The Eastman Dry Plate Company was born. More tinkering followed. In 1884, Eastman invented roll film; four years later he came up with a camera capable of taking advantage of that roll. In 1888, that camera became commercially available, later marketed under the slogan "You press the button, we do the rest."[2] The Eastman Dry Plate Company had become the Eastman Company, but that name wasn't quite catchy enough. Eastman wanted something stickier, something that people would remember and talk about. One of his favorite letters was K. In 1892, the Eastman Kodak Company was born.

In those early years, if you would have asked George Eastman about Kodak's business model, he would have said the company was somewhere between a chemical supply house and a dry goods purveyor (if dry plates can be considered dry goods). But that changed quickly. "The idea gradually dawned on me," Eastman said, "that what we were doing was not merely making dry plates, but that we started out to make photography an everyday affair."[3] Or, as Eastman later rephrased it, he wanted to make photography "as convenient as a pencil."

And for the next hundred years, Eastman Kodak did just that.

The Memory Business

Steven Sasson is a tall man with a lantern jaw. In 1973, he was a freshly minted graduate of the Rensselaer Polytechnic Institute. His degree in electrical engineering led to a job with Kodak's Apparatus Division research lab, where, a few months into his employment, Sasson's supervisor, Gareth Lloyd, approached him with a "small" request. Fairchild Semiconductor had just invented the first "charge-coupled device" (or CCD)—an easy way to move an electronic charge around a transistor—and Kodak needed to know if these devices could be used for imaging.[4] Could they ever.

By 1975, working with a small team of talented technicians, Sasson used CCDs to create the world's first digital still camera and digital recording device. Looking, as *Fast Company* once explained, "like a '70s Polaroid crossed with a Speak-and-Spell,"[5] the camera was the size of a toaster, weighed in at 8.5 pounds, had a resolution of 0.01 megapixel, and took up to thirty black-and-white digital images—a number chosen because it fell between twenty-four and thirty-six and was thus in alignment with the exposures available in Kodak's roll film. It also stored shots on the only permanent storage device available back then—a cassette tape. Still, it was an astounding achievement and an incredible learning experience.

Portrait of Steven Sasson with first digital camera, 2009

Source: *Harvey Wang*, From Darkroom to Daylight

"When you demonstrate such a system," Sasson later said, "that is, taking pictures without film and showing them on an electronic screen without printing them on paper, inside a company like Kodak in 1976, you have to get ready for a lot of questions. I thought people would ask me questions about the technology: How'd you do this? How'd you make that work? I didn't get any of that. They asked me when it was going to be ready for prime time? When is it going to be realistic to use this? Why would anybody want to look at their pictures on an electronic screen?"[6]

In 1996, twenty years after this meeting took place, Kodak had 140,000 employees and a $28 billion market cap. They were effectively a category monopoly. In the United States, they controlled 90 percent of the film market and 85 percent of the camera market.[7] But they had forgotten their business model. Kodak had started out in the chemistry and paper goods business, for sure, but they came to dominance by being in the convenience business.

Even that doesn't go far enough. There is still the question of what exactly Kodak was making more convenient. Was it just photography? Not even close. Photography was simply the medium of expression— but what was being expressed? The "Kodak Moment," of course— our desire to document our lives, to capture the fleeting, to record the ephemeral. Kodak was in the business of recording memories. And what made recording memories more convenient than a digital camera?

But that wasn't how the Kodak Corporation of the late twentieth century saw it. They thought that the digital camera would undercut their chemical business and photographic paper business, essentially forcing the company into competing against itself. So they buried the technology. Nor did the executives understand how a low-resolution 0.01 megapixel image camera could hop on an exponential growth curve and eventually provide high-resolution images. So they ignored it. Instead of using their weighty position to corner the market, they were instead cornered by the market.

Do the Math

Back in 1976, when Steven Sasson first demonstrated the digital camera at Kodak, he was immediately asked for a ready-for-prime-time estimate. How long, frightened executives wanted to know, until his new invention posed a serious threat to the company's market dominance? Fifteen to twenty years, Sasson said.[8]

In arriving at this answer, Sasson made a quick estimation and did a quick calculation. He estimated the number of megapixels that would satisfy an average consumer at two million. Then, in order to figure out the time it would take for these two million megapixels to become

commercially available, Sasson relied on Moore's law for his calculation—and that's where the trouble started.

In 1965, Gordon Moore, the founder of Intel, noticed the number
of integrated circuits on a transistor had been doubling every twelve to
twenty-four months. The trend had been going on for about a decade
and, Moore predicted, would probably last for another.[9] About this
last part, he was off by a bit. All told, Moore's law has held steady for
nearly sixty years. This relentless progress in price and performance is
the reason the smartphone in your pocket is a thousand times faster
and million times cheaper than a supercomputer from the 1970s. It is
exponential growth in action.

Unlike the +1 progression of linear growth, wherein 1 becomes 2
becomes 3 becomes 4 and so forth, exponential growth is a compound
doubling: 1 becomes 2 becomes 4 becomes 8 and so on. And this is the
problem: This doubling is unusually deceptive. If I take 30 large linear steps (say three feet, or one meter per step) from my Santa Monica
living room, I end up 30 meters away, or roughly across the street. If,
alternatively, I take 30 exponential steps from the same starting point,
I end up a billion meters away, or orbiting the Earth 26 times. And
this was exactly where Kodak went wrong—they underestimated the
power of exponentials.

The Six Ds

Underestimating the power of exponentials is easy to do. We hominids
evolved in a world that was local and linear. Back then, life was local
because everything in our forebears' lives was usually within a day's walk.
If something happened on the other side of the planet, we knew nothing
about it. Life was also linear, meaning nothing changed over centuries or
even millennia. In stark contrast, today we live in a world that is global
and exponential. The problem is that our brains—and thus our percep-
tual capabilities—were never designed to process at either this scale or
this speed. Our linear mind literally cannot grok exponential progression.

But if the goal is to avoid Kodak's errors (if you're a company) or to exploit Kodak's errors (if you're an entrepreneur), then you need to have a better understanding of how this change unfolds—and that means understanding the hallmark characteristics of exponentials. To teach these, I have developed a framework called the Six Ds of Exponentials: digitalization, deception, disruption, demonetization, dematerialization, and democratization. These Six Ds are a chain reaction of technological progression, a road map of rapid development that always leads to enormous upheaval and opportunity.

So let's follow the chain reaction.

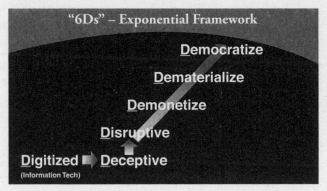

The 6 Ds of Exponentials: Digitalization, Deception, Disruption, Demonetization, Dematerialization, and Democratization

Source: Peter H. Diamandis, www.abundancehub.com

Digitalization. This idea starts with the fact that culture makes progress cumulative. Innovation occurs as humans share and exchange ideas. I build on your idea; you build on mine. This type of exchange was slow in the early days of our species (when all we had as a means of transmission was storytelling around the campfire), picked up with the printing press, then exploded with the digital representation, storage, and exchange of ideas made possible by computers. Anything that could be digitized— that is, represented by ones and zeros—could spread at the speed of light (or at least the speed of the Internet) and became free to reproduce and share. Moreover, this spreading followed a consistent pattern: an exponential growth curve. In Kodak's case, once the memory business went from a physical process (that is to say, imaged on film, stored on paper) to a digital process (imaged and stored as ones and zeros), its growth rate became entirely predictable. It was now on an exponential curve.

Of course, it's not just Kodak. Anything that becomes digitized (biology, medicine, manufacturing, and so forth) hops on Moore's law of increasing computational power.[10] Thus the first of our Ds is *digitalization*, for the simple reason that once a process or product transitions from physical to digital, it becomes exponentially empowered.

Deception. What follows digitalization is deception, a period during which exponential growth goes mostly unnoticed. This happens because the doubling of small numbers often produces results so minuscule they are often mistaken for the plodder's progress of linear growth. Imagine Kodak's first digital camera with 0.01 megapixels doubling to 0.02, 0.02 to 0.04, 0.04 to 0.08. To the casual observer, these numbers all look like zero. Yet big change is on the horizon. Once these doublings break the whole-number barrier (become 1, 2, 4, 8, etc.), they are only twenty doublings away from a millionfold improvement, and only thirty doublings away from a billionfold improvement. It is at this stage that exponential growth, initially deceptive, starts becoming visibly disruptive.

Disruption. In simple terms, a disruptive technology is any innovation that creates a new market and disrupts an existing one. Unfortunately, as disruption always follows deception, the original technological threat often seems laughably insignificant. Take the first digital camera. Kodak took great pride in things like convenience and image fidelity. Neither were present in Sasson's original offering. His camera took twenty-three seconds to snap and store a 0.01 megapixel, black-and-white photograph. Well, no threat there.

In the eyes of the Kodak brass, Sasson's innovation would remain more toy than tool for many years to come. With their focus on the quarterly profits of their chemicals and paper business, they didn't understand the disruption soon to be wrought by exponentials. If Kodak had done the math, their executives would have realized that the desire to not compete against themselves was actually a decision to put themselves out of business.

And out of business is where the company went. By the time Kodak realized its error, it was unable to keep pace with the digitalization of the industry. Kodak began to struggle in the nineties and stopped turning a profit by 2007, then filed for Chapter 11 in January of 2012.[11] Because it forgot its mission and failed to do the math, a gargantuan hundred-plus-year-old industry foundered and became yet another cautionary tale about the disruptive nature of exponential growth.

We live in an exponential era. This kind of disruption is a constant. For anyone running a business—and this goes for both start-ups and legacy companies—the options are few: Either disrupt yourself or be disrupted by someone else.

The Last Three Ds

Digitalization, deception, and disruption have radically reshaped our world, but the chain reaction we're tracking is cumulative. Thus the three Ds that follow—demonetization, dematerialization, and democratization—are far more potent than their predecessors.

Demonetization. This means the removal of money from the equation. Consider Kodak. Their legacy business evaporated when people stopped buying film. Who needs film when there are megapixels? Suddenly one of Kodak's once-unassailable revenue streams came free of charge with any digital camera.

In one sense, this transformation is the downstream version of what former *Wired* editor-in-chief Chris Anderson meant in his book *Free*. In *Free*, Anderson argues that in today's economy one of the easiest ways to make money is to give stuff away.[12] Here's how he explains it:

I'm typing these words on a $250 "netbook" computer, which is the fastest growing new category of laptop. The operating system happens to be a version of free Linux, although it doesn't matter since I don't run any programs but the free Firefox Web browser. I'm not using Microsoft Word, but rather free Google Docs, which has the advantage of making drafts available to me wherever I am, and I don't have to worry about backing them up since Google takes care of that for me. Everything else I do on this computer is free, from my email to my Twitter feeds. Even the wireless access is free, thanks to the coffee shop I'm sitting in.

And yet Google is one of the most profitable companies in America, the Linux ecosystem is a $30 billion industry, and the coffee shop seems to be selling $3 lattes as fast as they can make them.

Billions and billions in goods and services, as Anderson pointed out, are now changing hands sans cost. Now, sure, there is *loss-leader free*—as with Google's giving away their browser but making a killing off the information they gather along the way—and there's open-source efforts like Wikipedia, Linux, and all the rest, which are *actually free*. Either way, it's a shadow economy, yet happening in plain sight. Literally. At the time Anderson wrote *Free*, beyond a few extremely obscure papers, economists had not studied the idea of free in the marketplace. It was a blank spot on the map. In other words, even people who make their living studying economic trends were fooled. Once demonetization arrived, they didn't know what hit them.

Nor is it just economists or, for that matter, Kodak executives. Skype demonetized long-distance telephony; Craigslist demonetized classified advertising; Napster demonetized the music industry. This list goes on and on. More critically, because demonetization is also deceptive, almost no one within those industries was prepared for such radical change.

Dematerialization. While demonetization describes the vanishing of the money once paid for goods and services, dematerialization is about the vanishing of the goods and services themselves. In Kodak's case, their woes didn't end with the vanishing of film. Following the invention of the digital camera came the invention of the smartphone—which soon came standard with a high-quality, multi-megapixel camera. Poof! Now you see it; now you don't. Once those smartphones hit the market, the digital camera itself dematerialized. Not only did it come free with most phones, consumers expected it to come free with most phones. In 1976, Kodak controlled 85 percent of the camera business. By 2008—one year after the introduction of the first iPhone (the first smart phone with a high-quality digital camera)—that market no longer existed.

How Many Photos Are Taken Each Year?

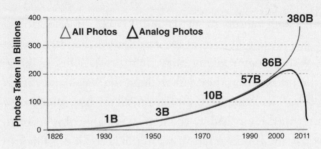

The decline of print and explosion of digital photography

Source: http://digital-photography-school.com/history-photography

What makes this story even stranger is that Kodak knew this change was coming. Moore's law was well established at that point, already driving the ceaseless expansion of memory storage capacity, the process that would lead to the demonetization of photography. Kodak's engineers surely knew this. They arguably also knew about Hendy's law—which states that the number of pixels per dollar found in digital cameras doubles every year—as the term was coined by an employee of Kodak Australia, Barry Hendy. The writing wasn't just on the wall for Kodak—they had put it there themselves. Yet Kodak still failed to stay ahead of this curve.

Take a look at the chart below.

>$900,000 worth of applications in a smart phone today

Application	$ (2011)	Original Device Name	Year*	MSRP*	2011's $
1 Video conferencing	free	Compression Labs VC	1982	$250,000	$586,904
2 GPS	free	TI NAVSTAR	1982	$119,900	$279,366
3 Digital voice recorder	free	SONY PCM	1978	$2,500	$8,687
4 Digital watch	free	Seiko 035Q Astron	1969	$1,250	$7,716
5 5 Mpixel camera	free	Canon RC-701	1986	$3,000	$6,201
6 Medical library	free	e.g. CONSULTANT	1987	Up to $2,000	$3,988
7 Video player	free	Toshiba V-8000	1981	$1,245	$3,103
8 Video camera	free	RCA CC010	1981	$1,050	$2,617
9 Music player	free	Sony CDP-101 CD player	1982	$900	$2,113
10 Encyclopedia	free	Compton's CD Encyclopedia	1989	$750	$1,370
11 Videogame console	free	Atari 2600	1977	$199	$744
Total	free				$902,809

*Year of Launch

The roughly $900,000 worth of applications in a smart phone today

Source: Abundance: The Future Is Better Than You Think, page 289

* Manufacturer's Suggested Retail Price

It shows all the 1980s luxury technologies that have dematerialized and now come standard with your average smartphone. An HD video camera, two-way video conferencing (via Skype), GPS, libraries of books, your record collection, a flashlight, an EKG, a full videogame arcade, a tape recorder, maps, a calculator, a clock . . . just to name a few. Thirty years ago the devices in this collection would have cost hundreds of thousands of dollars; today they come free or as apps on your phone. And smartphones are the fastest-spreading technology in humanity's history.

Democratization. Obviously, this chain of vanishing returns has to end somewhere. Sure, film and cameras now come free with smartphones, but there are still the hard costs of the phone with which to contend. Democratization is what happens when those hard costs drop so low they becomes available and affordable to just about everyone. To put this in perspective, let's return to Kodak.

The company didn't just make money selling cameras and selling film, they also sold everything on the back end of the process: they developed the film, manufactured the paper the photographs were printed on, and manufactured the chemicals used to develop that film. Why was this such a good business? First, when you snapped your photos, you had no idea which of them would actually turn out to be any good, so you printed them all. Remember those rolls of film where nothing was in focus? You still paid. Second, snapping photos was only part of the fun; printing extra copies and sharing those photos was the real treat.

Two decades back, the only people who could snap and share at will were those wealthy enough to afford the considerable paper, printing, and processing costs associated with several thousand photographs. But with the digital camera, you gained the benefit of knowing in advance which shots are actually worth printing, and with the creation of photo-sharing websites like Flickr, you could avoid printing altogether. The sharing of images became free, fast, and completely democratized.

Democratization is the end of our exponential chain reaction, the logical result of demonetization and dematerialization. It is what happens when physical objects are turned into bits and then hosted on a digital platform in such high volume that their price approaches zero. Such is the case with today's smartphones and tablets. In fact, it's also the case with wireless connectivity, which is what allows these devices to communicate with the Internet. Right now, Google and Facebook are in an arms race, with plans to spend billions to launch drones, balloons, and satellites capable of providing free or ultra-low-cost Internet access to every human on Earth.[13]

Many legacy institutions (like Kodak) once were able to make a great living resting on their laurels. According to Yale professor Richard Foster, in the 1920s the average life span of an S&P 500 company was sixty-seven years.[14] Not anymore. Today the final three Ds in our chain reaction can disassemble companies and disrupt industries almost overnight, reducing the average life span of a twenty-first-century S&P 500 company to only fifteen years. Ten years from now, according to research done at the Babson School of Business, more than 40 percent of today's top companies will no longer exist.[15] "By 2020," comments Foster, "more than three quarters of the S&P 500 will be companies that we have not heard of yet."[16]

For linear-thinking companies, the six Ds of exponentials are the six horsemen of the apocalypse—no question about it. But this is not a book designed to protect legacies from exponentials. It is a book for entrepreneurs looking to harness the power of exponentials to start building new, bold legacies. For these exponential entrepreneurs, the future is not about disruptive stress; rather, it's frothing with disruptive opportunity.

The New Kodak Moment

In his book *Exponential Organizations*, Singularity University global ambassador and former head of innovation at Yahoo Salim Ismail defines an *exponential organization* as one whose impact (or output)— because of its use of networks or automation and/or its leveraging of the crowd—is disproportionally large compared to its number of employees.[17] A *linear organization*—like, say, Kodak—is the opposite: lots of employees and lots of physical processes and facilities. For all of the twentieth century, exponential organizations did not exist and linear companies were protected from upstart intruders by sheer size. Those days are gone.

In October 2010, a couple of young Stanford grads, Kevin Systrom and Mike Krieger, founded an exponential organization called Instagram. *Wired* magazine described Instagram as a "Shiva-the-destroyer application posing as a hipster hobby."[18] And what was that hobby exactly? The next step in George Eastman's vision of making photography—to borrow the phrase—as convenient as a pencil.

In this, Instagram was extraordinary. Combined with the explosion of high-resolution multi-megapixel smartphone cameras, this renegade start-up completely demonetized, dematerialized, and democratized the capturing and sharing of photographic memories. Sixteen months after the founding of the company, Instagram was valued at $25 million.

Instagram Number of Users

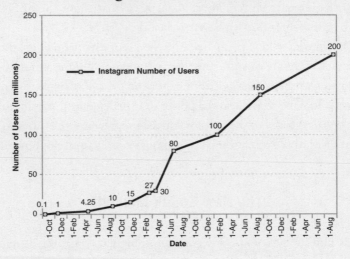

Sources: *http://instagram.com/press;*

http://www.macstories.net/news/instagrams-rise-to-30-million-users-visualized/

In April 2012, Instagram for Android was released. Downloaded more than a million times in one day, it was the killer app for the already killer company.[19] Instagram's value shot up to $500 million. Enter Facebook.

Facebook is also in the life-sharing, life-documenting business—and they did the math. Instagram was growing exponentially. With nearly 30 million users, it wasn't just a photo-sharing service; it had become *the* photo-sharing service, with a very powerful social network to boot. Facebook didn't want the competition, and they didn't want to play catch-up. Thus, on April 9, 2012, just three months after Kodak filed for bankruptcy, Instagram and its thirteen employees were bought by Facebook for $1 billion.[20]

But how is this possible? How did Kodak—a hundred-year-old behemoth with 140,000 employees and a 1996 value of $28 billion—fail to take advantage of the most important photographic technology since roll film and end up in bankruptcy court? Simultaneously, how did a handful of entrepreneurs working out of the proverbial Silicon

Valley garage go from start-up to a billion-dollar buyout in eighteen months, with a little more than a dozen employees? Simple: Instagram was an exponential organization.

Welcome to the New Kodak Moment—the moment when an exponential force puts a linear company out of business. As we shall see over and over again, these New Kodak Moments are not aberrations. Rather, they are the inevitable result of the six Ds of exponential growth. And for those linear-thinking executives trying to hang on to their jobs, this leads us to three final Ds: distraught, depressed, and departed. But for exponential entrepreneurs, these New Kodak Moments are ripe with possibility.

A Question of Scale

Today, exponential technology is not just putting linear companies out of business, it's also putting linear industries out of business. It's shifting the entire landscape, disrupting traditional industrial processes—like the process by which consumer goods are invented and come to market. For the right entrepreneur, there's considerable opportunity within this disruption.

Ben Kaufman was the right entrepreneur.[21]

Ben Kaufman was born in 1986 and raised on Long Island in New York. He was a horrible student, but also a horribly inventive student. In his senior year of high school, Kaufman decided he wanted to build a "stealth iPod"—a device that would allow him to listen to his iPod shuffle, in class and in secret, without his teacher ever noticing.

So Kaufman came home from school and built a prototype out of spare parts—mainly ribbon and gift wrap paper—proving to himself that the design would work. He also felt that other people would want one as well. But rather than be satisfied with a prototype, he somehow convinced his parents to remortgage their home and lend him $185,000 to take his invention to market. With cash in hand, Kaufman was on the next flight to China.

Once he got to China, Kaufman learned the hard way that creating a consumer product wasn't just about raising the money. "You need access to industrial design, distribution, marketing, branding, packaging . . . There's literally a list of thirty things you need to have to be successful . . . It's just *really, really* hard."

Kaufman persevered. He founded Mophie, an Apple accessories company, and brought that initial product to market. Then, using his hard-won skills, his company delivered several dozen other Apple accessories. After that came Kaufman's next company, Quirky, the inspiration for which came to him early one morning in New York.

"I was sitting on the subway," he explains, "and there was a woman wearing my first product, the stealth iPod I had prototyped back in high school. Seeing that made me realized that I wasn't unique in having a good idea. What was unique were all the circumstances that lined up in order for me to execute on my idea. It hit me that it wasn't just me. Invention is typically inaccessible. It's really, really hard for everyone."

Standing in the way of invention is financing, engineering, distribution, and legalities—all the myriad quagmires that we loosely call the process of product development. So, in Quirky, Kaufman created a company whose mission is to "make invention accessible." Or as he says: "Make it possible for all people regardless of their love, circumstances, and pedigree to execute on their great ideas."

To do just that, Kaufman swapped out the linear for the exponential, open-sourcing the entire *process* of product development. A Quirky user simply posts his or her product idea to the site, where other users vote on its feasibility and desirability. And what the crowd likes, the crowd builds, one crowdsourced, open-sourced step at a time. This means the Quirky community will shepherd your idea from prototype onto the shelves at Target, while sidestepping all of the traditional development bottlenecks. That's also where the name Quirky comes from. "It's a weird way of looking at product development," explains Kaufman. "We're changing the way that communities interact and what they do online together. Now they're not just finding each other and sharing things, they're actually building things."

Quirky launched in 2009, quickly raised over $79 million in funding, and has already introduced several hundred products to market.[22] There's a flexible power strip called Pivot Power, a collapsible clothes hanger called Solo. There are new bookcases, backpacks, cord management devices, cleaning products, cooking products, and just about anything else you can imagine. But the real difference is speed. Go back twenty-five years and the time it took for any of these inventions to come to market was measured in years. With Quirky, it takes about four months.

And unlike linear companies, whose old-world structure and processes limit their ability to rapidly introduce new products, Quirky, with a community north of 800,000, has already released over three hundred products and is currently introducing two to three new ones a week. Rather than closed-door design sessions and behind-the-scenes marketing moves, everything at Quirky is transparent, available online, and open to the public. This is to say, everything at Quirky is designed to let any entrepreneur take advantage of the amazing power of exponential organizational tools such as crowdsourcing.

And the entrepreneur should take advantage. The goal here is not to teach you how to become Ben Kaufman, it's to teach you to harness exponential platforms like Quirky, or to encourage you to create similar platforms yourself.

Consider Candace Klein, a crowdsourcing expert and the very busy CEO of Bad Girl Ventures, a company that helps women start businesses. Every Saturday night, Klein gets together with a group of women friends for cocktails. "Some of us run businesses," explains Klein, "and some of us are stay-at-home moms, but we're all really inventive and entrepreneurial. We usually spend Saturday night talking about whatever it is we'd like to invent next. And we park these ideas on Quirky. Sometimes that takes a little work, but most of the time we're done putting the idea onto the site in about fifteen minutes."[23]

Over the past few years, the ideas that Candace and her cocktail klatch have parked on Quirky have generated over $100,000 in revenue, a six-figure salary for work done while getting buzzed.

But that's not the whole story. Equally important is that Kaufman's success and, by extension, Klein's success, rest upon another radical shift in the playing field—a shift in scale.

In the early days of exponentials, disruptions were of the Kodak variety. Companies that made digitizable goods and services—the publishing business, the music business, the memory business, etc.—were threatened. But Quirky gives us a look at the next level up. It is no longer goods and services being subjected to the Six Ds; it's whole industrial processes. Quirky is an alternative to the entire twentieth-century product development chain—an alternative to every single step in that once hugely capital-intensive process.

And again, it's not just Quirky. Go back ten years, and hospitality and lodging was an incredibly capital-intensive business. If you wanted to build a nationwide chain of available hotel rooms you had to, well, build those actual hotel rooms. But that's not what Airbnb did.

Technically, Airbnb is a hosting platform, except that term doesn't exactly reflect the scale of disruption the company has wrought. By providing a place to post available spare bedrooms, open garage apartments, even empty vacation homes, this site allows anyone to turn unused space into a bed-and-breakfast. By mid-2014, just six years into their existence, Airbnb had over 600,000 listings in 34,000 cities and 192 countries and had served over 11 million guests. Most recently the company was valued at $10 billion—making it worth more than Hyatt Hotels Corporation ($8.4 billion)—and all without building a single structure.[24]

Then there's Uber, a different kind of hosting platform—one going head-to-head with the taxi and limousine industry.[25] Download the Uber app and you can order a car, get information about the driver, watch the car's approach on a map, and, with your credit card already stored online, pay instantly. Yet Uber doesn't own a fleet of vehicles or manage a stable of drivers. The company simply provides a connection between people with assets (aka luxury cars) and you, the customer. In other words, by putting would-be passengers together with luxury vehicle owners, Uber cut out the middleman, dematerialized a boatload of infrastructure, and democratized a sizable segment of the transportation industry. And fast. Four years after launching their mobile, Uber is operational in thirty-five cities, and worth $18 billion.

Quirky, Airbnb, and Uber are great examples of entrepreneurs taking advantage of the expanding scale of exponential impact. They have created billion-dollar companies in record time. They are the absolute inverse of everything we believed was true about scaling up a capital-intensive businesses. For most of the twentieth century, scaling up such businesses required massive investments and time. Adding workforce, constructing buildings, developing vastly new product suites— no wonder implementation strategies stretched years into decades. It wasn't unusual for a board of directors to "bet the company" on a new and extremely expensive direction whose outcome would remain unknown until long after most of those board members retired.

That was then.

Today linear organizations are at dire risk from the Six Ds, but exponential entrepreneurs have never had it so good. Today the shift from "I've got a neat idea" to "I run a billion-dollar company" is occurring faster than ever.

This is possible, in part, because the structure of exponential organizations is very different. Rather than utilize armies of employees or large physical plants, twenty-first-century start-ups are smaller organizations focused on information technologies, dematerializing the once physical and creating new products and revenue streams in months, sometimes weeks. As a result, these lean start-ups are the small furry mammals competing with the large dinosaurs—meaning they're one asteroid strike away from world dominance.

Exponential technology is that asteroid.

In times of dramatic change, the large and slow cannot compete with the small and nimble. But being small and nimble requires a whole lot more than just understanding the Six Ds of exponentials and their expanding scale of impact. You'll also need to understand the technologies and tools driving this change. These include exponential technologies like infinite computing, sensors and networks,

3-D printing, artificial intelligence, robotics, and synthetic biology and exponential organizational tools such as crowdfunding, crowdsourcing, incentive competitions, and the potency of a properly built community. These exponential advantages empower entrepreneurs like never before.

Welcome to the age of exponentials.

CHAPTER TWO

Exponential Technology

The Democratization of the Power to Change the World

Reading Exponential Road Maps

In his book *The Prime Movers,*[1] psychologist Edwin Locke identifies the core mental traits of great business leaders—Steve Jobs, Sam Walton, Jack Welch, Bill Gates, Walt Disney, and J. P. Morgan, to name only a few. While a number of variables contributed to their success, Locke found one key trait they all shared: vision.

"It's the ability to see ahead that truly set each of these men apart," says Locke.[2] "The data shows that companies consistently fail when they rest on their laurels and think that what worked yesterday will work today or tomorrow. Great leaders all have the ability to see farther and the confidence to drag their organizations toward that vision. Look at Steve Jobs. He wasn't very nice about it, but if you told Jobs something was impossible, all he would do was disagree and walk away. He had no time for impossible. He had a vision of the future and would not be swayed."

Thus the goal of this chapter and the next is to give you exactly what these great leaders had: the vision not to be swayed. This means giving you a much firmer sense of the future, and this requires a three-step process. First, in the next section, we're going to revisit the deceptive nature of exponentials and point out a few indicators that typically mark the transition from deceptive to disruptive—which is exactly when entrepreneurs need to insert themselves into the equation. Next, we're going to see how all this plays out in the real world, drilling down into the past, present, and future of 3-D printing—a technology currently transitioning from deceptive to disruptive and ripe with possibility. To explore these possibilities, we'll meet a visionary leader in the field and a couple of entrepreneurs perhaps not unlike yourself, folks who had little knowledge of this technology yet still figured out a way to harness its power and start companies very capable of causing billon-dollar disruptions.

Then, in the next chapter, we'll block out a few more exponentially advancing technologies, seeing the future potential for networks and sensors, infinite computing, artificial intelligence, robotics, and synthetic biology. While these technologies are already producing disruptive growth and have begun transforming our world, the "expert only" nature of their interfaces (more on this in a moment) and their stratospheric price tags have kept them primarily in the hands of billion-dollar companies (think Google's use of artificial intelligence or Tesla Motors's use of robotics). But not for long. Decreasing prices, increasing performance, and the development of far friendlier user interfaces are making these platforms available to those with a clear vision of where they want to go. Thus each of these technologies hover on the verge of widespread adoption, and for those exponential entrepreneurs able to stay ahead of this curve, the opportunities are considerable.

The Hype Curve and the User Interface

If we want to stay ahead of this curve, it helps to understand a little more about the nature of exponential deception. That starts with understanding the powerful biases that inform the Gartner Hype Cycle (see below).

After a novel technology is introduced and begins gaining momentum, we tend to envision it in its final form—seriously overinflating our expectations for both its developmental timetable and its short-term potential. Invariably, when these technologies fail to live up to the initial hype—usually in that gap between deception and disruption on our list of the Six Ds—public sentiment for the technology falls into the trough of disillusionment. And this is where a great many of the technologies discussed in chapter 3 now sit. But when technologies are in the trough, we are again swayed by the hype (this time, the negative hype) and consistently fail to believe they'll ever emerge, thus missing their massively transformative potential.

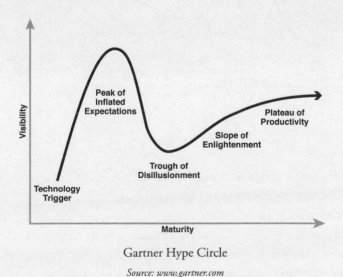

Gartner Hype Circle

Source: www.gartner.com

Take the personal computer. Back in the late 1960s, when folks like writer Stewart Brand (who coined the term *personal computer*) first started discussing the idea of the PC, it was with an incredible amount of "change the world" fervor.[3] Then the machines actually arrived, and all most people could do was play Pong. This was the trough of disillusionment, cultural deception at its finest. But imagine being able to take your knowledge of what computers can do today back to the early 1980s—what bold entrepreneurial business opportunities might this have unlocked for you?

Hype Cycle Indicators

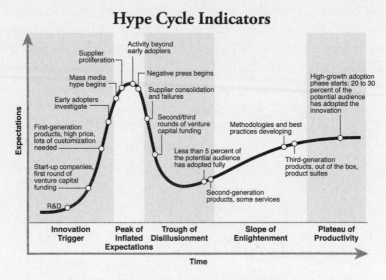

Gartner Hype Cycle Indicators

Source: www.gartner.com

Recognizing when a technology is exiting the trough of disillusionment and beginning to rise up the slope of enlightenment is critical for entrepreneurs. Reading an exponential curve like a road map, experts watch for a number of indicators—the development of best practices, supplier proliferation, secondary financings, among others. But for me, the most important telltale factor is the development of a simple and elegant user interface—a gateway of effortless interaction that plucks a technology from the hands of the geeks and deposits it with the entrepreneurs. In fact, it was exactly this kind of interface that transformed the Internet.

The Internet was born of frustration. In the early 1960s, researchers were titillated by computational possibility yet stymied by geography. Back then, there were only a few major computing centers on the planet. All those researchers who didn't happen to work at MIT or Caltech—well, they were just out of luck. Then, in April 1963, a computer scientist named J. C. R. Licklider wrote a memo to his colleagues proposing an "Intergalactic Computer Network"—a network that replaced traditional circuit-switching technology with the then new development of packet switching, allowing any researcher with a terminal and a phone line to connect to one of the computing centers they so desperately needed.[4] This was the birth of the Advanced Research Projects Agency Network (ARPANET), the foundational network that has since become today's Internet.

ARPANET became operational in 1975. It was mostly text-based, fairly complicated to navigate, and used primarily by scientists. All of this changed in 1993, when Marc Andreessen, a twenty-two-year-old undergraduate student at the University of Illinois, Urbana-Champaign, coauthored Mosaic—both the very first web browser and the Internet's first user-friendly user interface.[5] Mosaic unlocked the Internet. By adding in graphics and replacing Unix with Windows—the operating system that was then running nearly 80 percent of the computers in the world—Andreessen mainstreamed a technology developed for scientists, engineers, and the military. As a result, a worldwide grand total of twenty-six websites in early 1993 mushroomed into more than 10,000 sites by August 1995, then exploded into several million by the end of 1998.[6]

Industries Being Disrupted by 3-D Printing

Industries	Current Applications	Potential Future Applications
COMMERCIAL AEROSPACE AND DEFENSE[17]	• Concept modeling and prototyping • Structural and non-structural production parts • Low-volume replacement parts	• Embedding additively manufactured electronics directly on parts • Complex engine parts • Aircraft wing components • Other structural aircraft components
SPACE	• Specialized parts for space exploration • Structures using light-weight, high-strength materials	• On-demand parts/spares in space • Large structures directly created in space, thus circumventing launch vehicle size limitations
AUTOMOTIVE[18]	• Rapid prototyping and manufacturing of end-use auto parts • Parts and assemblies for antique cars and racecars • Quick production of parts or entire	• Sophisticated auto components • Auto components designed through crowdsourcing
HEALTH CARE[19]	• Prostheses and implants • Medical instruments and models • Hearing aids and dental implants	• Developing organs for transplants • Large-scale pharmaceutical production • Developing human tissues for regenerative therapies
CONSUMER PRODUCTS/RETAIL	• Rapid prototyping • Creating and testing design iterations • Customized jewelry and watches • Limited product customization	• Co-designing and creating with customers • Customized living spaces • Growing mass customization of consumer products

Sources: Deloitte analysis; CSC, 3D printing and the future of manufacturing, 2012

Graphic: Deloitte University Press | DUPress.com

This is the power of an elegant and robust user interface. It's also a sign that it's time for an exponential entrepreneur to get in the game. Certainly, deciding when a technology is ripe for entrepreneurial development is not too different from a venture capitalist deciding a technology is ripe for investment. Yet venture capitalists have dozens and dozens of theories about when a technology is actually ripe for investment, so why have I chosen this one indicator above all others? Simple. The creation of a simple and elegant user interface gives entrepreneurs the ability to harness this new tool to solve problems, start businesses, and most importantly, experiment. Think of the explosion of apps that followed Apple's creation of the App Store. As new entrepreneurs are constantly improving this new interface, they are also further enabling new entrepreneurs—meaning a positive feedback loop of increasing interface innovation develops. It's a virtuous cycle seen over and over again.

More exciting, it's exactly these kinds of robust, elegant inter-faces that are beginning to show up in half a dozen exponential technologies—meaning there are literally a half dozen Internet-sized opportunities becoming available to the clued-in entrepreneur.

3-D Printing: The Origins and Power of Additive Manufacturing

One such opportunity lies with 3-D printing, a technology now emerging from a thirty-year period of deceptive growth and beginning to disrupt a portion of the $10 trillion global manufacturing industry.[7] In the rest of this chapter, we're going to explore this technology's past, present, and future and then acquaint you with a few entrepreneurs pioneering that future. The goal here is both to familiarize you with this technology and use it as a real-time template for the Six Ds, exploring how select entrepreneurs have correctly read the cycle of hype and positioned themselves to take full advantage of this tech's exponential opportunity.

To accomplish this goal, we need to start at the beginning. So let's crank up the wayback machine and take a trip some 2.6 million years into our past, to what is now Southern Ethiopia, where one of our craftier ancestors picked up a pair of rocks and used one to chip away at the other until all that remained of the second was a sharp stone flake.[8] Eureka!

Stone chipping was the birth of tool use, but it was also the birth of *subtractive manufacturing*, a process of object creation wherein a larger block of material (i.e., a big flat stone) is subtracted from until all that remains is a large pile of debris and the desired object (i.e., the sharpened flake). And until recently, subtractive manufacturing was essentially how object creation got done.

Charles Hull changed this game. In the early eighties, Hull decided he wanted to help Detroit's ailing car industry compress their time to market and regain their competitive advantage. As he was then working for a small company in Southern California that specialized in developing applications for ultraviolet radiation, including curing (hardening) UV coatings and inks, Hull realized that curing's methodology opened the door for an entirely new manufacturing process. Instead of having to create new plastic parts and prototypes through subtractive methods, if he could figure out how to print sheets of UV-hardened plastic atop one another (and attach them to one another), he could build new automotive components via accretion—a method of *additive manufacturing* in which objects are built up one layer at a time. This was the birth of 3-D printing.[9]

To give you better idea of how this works, think of an ink-jet printer. These ubiquitous office products are 2-D printers that convert digital instructions (from your computer) into an "object" (aka printed text on a page) by printing along a two-dimensional (x and y) axis. A 3-D printer does the same except it adds in a vertical dimension (the z axis)—thus allowing for creation in all three dimensions.

Hull constructed his first 3-D printer in 1984, then founded the Valencia, California–based 3D Systems[10] to develop and commercialize the technology. Unfortunately, this was not easy. Over the next twenty years, development was slow (deceptive), incredibly expensive, and burdened with complicated user interfaces. All three factors prevented widespread adoption. By the early 2000s, despite their enormous first mover advantage, 3D Systems was on the verge of bankruptcy. "The company was a train wreck," says Avi Reichental.[11] "They had lost sight of the fact that their technology was accelerating exponentially. They had forgotten how to innovate."

And Reichental would know, as he was the person brought in to save the company.

On paper, Reichental was an odd choice for the job. Having spent the previous twenty-three years working for the Sealed Air Corporation, the inventors of Bubble Wrap, Reichental didn't know much about additive manufacturing. But what he did understand was innovation. "Sealed Air wasn't your standard package goods company," says Reichental. "It was more like a Silicon Valley start-up: totally entrepreneurial, always exploring new possibilities, always trying to crack open new markets."

As a result, Reichental worked dozens of different jobs during his Sealed Air tenure—eventually becoming the company's fourth-ranking officer and helping grow the firm from a 400-person, $100 million business (when he joined), into an 18,000-person, $5 billion behemoth (when he left). One of his many positions included a stint in the manufacturing department, where he was first introduced to 3-D printing as a way to speed up the prototyping process. This meant, when he first got the call about the 3D Systems job, Reichental had just enough information to do his due diligence. What he discovered was eye-opening.

"Sure," he says, "3D Systems was on the verge of extinction"—that is, still caught in its deceptive phase—"but I applied Moore's law to all the different verticals that went into this technology and saw they were all about to explode. By following out the exponential curves, I saw a technology that was about to transform how we create, what we create, and where we create. When you create additively, complexity comes free of charge. This means you're not constrained by any of the traditional manufacturing limits. You go straight from a computer file to a finished product, and can make millions of one-of-a-kind items without retooling or restocking. It's localized manufacturing at every scale. What I realized was that 3-D printing is a ubiquitous connection between the virtual and the actual; it's a technology with the potential to touch everything in our lives."

Reichental took the job. At the time, 3D Systems didn't have much of a product suite. They made six different kinds of printers—their most powerful device powered by two print engines—that could print in only four materials. Worse, 3D Systems didn't manufacture any of those materials, and only certain materials worked with certain machines.

Reichental's first order of business was to expand and integrate. "I wanted more printers, more materials to print with, and I wanted our printers to print with our materials so that you didn't have to be a super-expert to figure out how to work the machines. Simplicity was really important to me. I wanted people to expend their creativity thinking about new things to design, not about how to work the machines."

Along these same lines, Reichental also took to pestering Charles Hull. "Chuck still worked for the company. [Hull had retired, but was brought back as an interim CEO before Reichental took the job, and had stayed on in an advisory capacity.] His office was right by the coffee machine. One morning I stopped by and said: 'Our printers cost hundreds of thousands of dollars and you need to be astronaut-smart to run any of them. So, you know, why can't we make cheap push-button desktop versions?' He didn't treat me like a madman, but it was close. The next day I stopped by again and asked the same question. I did this every day for six weeks. Finally, one morning, he beat me to it. He came by my office and brought coffee. He had this incredible glow and said, 'I think I know how to do it.'"

And together they did. These days, 3D Systems is a thriving $6 billion[12] company making over forty different printers—the largest one capable of printing a Toyota Camry dashboard as a single piece. The simplest one is called the Cube, which costs $1,299 today (with plans to drop the price below $500 in the next couple of years). All told, these machines can print in over a hundred different materials, ranging from nylons, plastics, and rubbers all the way through biological materials (cells), real waxes, and even fully dense metals.

Yet, as was pointed out in the last section, an exponential technology doesn't really become disruptive until a powerful, user-friendly interface exists (think Mosaic). Thus 3D Systems has also expanded into software, with the goal of making their interface easy enough for children to use. They've been successful, too. "If you can point and click a mouse," says Reichental, "you can now design things for a 3-D printer. I call it the coloring-book model. In the past we had the canvas model. If you wanted to be a great artist, you had to have years of experience applying paint to a blank canvas. Now, with our coloring-book approach, if you want to be supercreative, all you have to know how to do is color between the lines."

What makes this development so much more important is that 3D

Systems isn't the only company designing new interfaces. Experimentation has begun, drawing in a multitude of other players. As a result, the field sits right about where the web sat when Marc Andreessen introduced Mosaic—completely primed for exponential explosion.

The Impact of Disruption

Even now, at the beginning of this explosion, the impact that 3-D printing is having on our world is considerable. Already the printing of standard consumer products—bowls, plates, smartphone cases, bottle openers, jewelry, and purses (made from mesh)—has gone from a hobby to a nascent industry. Dozens of websites now sell goods rendered with 3-D printers, and retailers are starting to get in on the action. As Mark Cotteleer,[13] Research Director at Deloitte Consulting, explains: "Our studies demonstrate two critical facts. First, break-even points for some objects, in particular smallish items made of plastic, can already surpass the hundred thousand unit mark—making them viable for many types of consumer products. Second, there is clear evidence that even for individual households, a consumer level additive manufacturing device can quickly manufacture enough goods to pay for itself, and thus represents an attractive financial investment for some US households."

On a larger scale, 3-D printing is making its presence felt in the transportation industry. Today most cars coming out of America, Europe, and Japan include 3-D printed parts. In September 2014, at the International Manufacturing Technology Show in Chicago, Local Motors CEO Jay Rogers (whom we will meet again later) and his team 3-D printed an entire car on site in one day.

Rogers describes digital manufacturing as the third industrial revolution. "The first revolution was the steam engine. Henry Ford gave us the second revolution, mass production, in which you can make something cheap as long as you make a million of them. The third revolution comes from the democratization of manufacturing, wherein a new car design does not require a new plant to be built."[14]

This third revolution is also impacting the aerospace industry. SpaceX recently announced it will 3-D print much of the rocket engine used in the Dragon 2 capsule,[15] Boeing currently 3-D prints over two hundred parts for ten different aircraft platforms,[16] and my own company, Planetary Resources (more on Planetary later), is 3-D printing much of the spacecraft that will travel to and prospect near-Earth asteroids.

And the financial impact of 3-D printing in the transportation industry cannot be overstated. CFM International's next generation superefficient LEAP airplane engine (expected commercially by 2016) uses 3-D printing to manufacture a radically new kind of fuel nozzle—impossible to manufacture with conventional machining processes—that reduces fuel use by 15 percent, a figure that, across the lifetime of a plane, translates into hundreds of billions of dollars of future savings.[17]

Medical devices are even further along. Because 3-D printing allows products to be perfectly matched to an individual's body shape, 3-D printers are being used today to make individually customized surgical tools, bone implants, prosthetic limbs, and orthodontic devices—all of which significantly enhance patient outcomes. It's also worth pointing out how fast this is happening.

In 2010, in *Abundance*, we reported on the work done by the incredibly talented Scott Summit, an industrial designer by trade, who was using 3-D printers to make customer-designed prosthetic limbs and back braces. Those medical devices were 3-D printed in a one-off fashion. Today, just three years later, Summit has joined 3D Systems, where he is helping take medical manufacturing to scale. Case in point, the company now provides the manufacturing infrastructure for every hearing aid device around the world, and over 95 percent of those are completely 3-D printed.

Another example of large-scale medically related 3-D printing can be found in the fully automated factories of Align Technology, the makers of Invisalign—the clear plastic teeth-straightening alternative to metal braces. This factory 3-D prints 65,000 distinct aligners every day. "Last year alone," says Reichental, "they printed seventeen million pairs of fully customized one-offs in a factory of the future not much bigger than a large college lecture hall."

Of course, the impact made by 3-D printing is going to stretch far further than just consumer goods and transportation and medical devices. Every aspect of the $10 trillion manufacturing sector has the potential to be transformed. That's $10 trillion worth of opportunity. So how can exponential entrepreneurs with little knowledge of this platform use it to disrupt industries and tackle the bold? Well, let's meet a few such innovators and find out.

Made in Space

I first met Aaron Kemmer, Michael Chen, and Jason Dunn, a trio of bold-minded innovators, during the summer of 2010 at the Singularity University Graduate Studies Program. It was a passion for space that brought them together. "That's why we came to SU," explains Kemmer.[18] "We were all serial entrepreneurs hunting for a big idea. We wanted to start a company that would help open the space frontier. We were definitely not thinking the way forward was going to be 3-D printing."

It was SU chair of robotics and three-time shuttle astronaut Dr. Dan Barry who pointed them in that direction. "We knew a little bit about 3-D printing," says Chen, "but only because Jason, with his aerospace background, had played with it a little in college. But when we were doing analysis—just looking at all the different exponential technologies and trying to come up with our idea—Dan Barry kept wandering over and telling us he had been to the International Space Station (ISS), and wow, having a 3-D printer on the ISS would sure be useful."

Eventually, they decided to figure out *how* useful.

"It's a supply chain problem," explains Dunn. "The ISS is at the back end of the longest, most complicated, and most expensive supply chain in existence. Launch costs are roughly ten thousand dollars a pound. And any object sent into space has to be durable enough to survive the eight minutes of high g-forces it takes to get out of the Earth's gravity well—which means building heavier objects. But any additional weight imposes a double penalty: Not only does every extra pound cost extra money, but it requires extra fuel to get off the planet, which means even more money."

Plus, when parts aboard the station break, resupply can take months and months. This is why there are over a billion manifest parts (meaning they've been paid for but have not necessarily flown yet) aboard the ISS. And after doing more research, Kemmer, Dunn, and Chen realized that 30 percent of these parts were plastic—meaning they should be printable with already available, off-the-shelf 3-D printing technology.

This was the birth of Made in Space, our first off-world 3-D printing company and a great example of exponential entrepreneurship. Their entire business model is based on exponential development curves. Their first offering, launched to the ISS in the fall of 2014, is the simplest: a 3-D printer that prints plastic parts.

In itself, this will bring on a manufacturing revolution of sorts. "The first 3-D printers on the ISS will be able to build objects that could never be manufactured on Earth," says Kemmer. "Imagine, for example, building a structure that couldn't withstand its own weight."

Following out the exponential curves a tiny bit further, Made in Space's next iteration is an advanced materials and multiple materials 3-D printer—which means that some time in the next five years 60 percent of the parts in use on the ISS will be printable. And just behind this version is the real game changer: a 3-D printer capable of printing electronics.

Consider the latest trend in satellite technology: CubeSats. These are tiny satellites weighing only a kilogram made in the shape of a ten-centimeter cube. They're so simple to build that almost anyone can pull it off (free instructions are available online), yet they can be deceptively powerful when deployed as a swarm, often taking the place of much bigger satellites. CubeSats themselves are cheap to make (about $5,000 to $8,000).[19] Launching them is the real expense (still tens of thousands of dollars). But that's today. If we wait a few more years, Made in Space can solve this problem for pennies on the dollar.

"Turns out," says Dunn, "the ISS [is] a perfect platform for launching things into low-Earth orbit. Already our printers can print the cube portion of a CubeSat, and we've also printed the electronics in our lab. It's hard to say for sure, but around 2025, we should be able to print electronics aboard the ISS. This means we'll be able to email hardware into space for free, rather than paying to have it launched there."

Of course, the big dream is to be able to create 3-D printers capable of printing entire space stations in space and, even better, to do it with materials mined from space. Once this becomes possible, the creation of legitimate off-world habitats (i.e., space colonies) becomes a viable reality.

"Imagine being able to colonize a distant planet by bringing nothing but a 3-D printer and some mining equipment," says Mike Chen. "It might sound like science fiction, but the first steps toward making it a reality are happening in our lab right now, and aboard the ISS."

What does this all mean? It means that while Made in Space started off disrupting a billion-dollar spare parts industry, the exponential growth curves that underpin their business model lead them directly toward first mover advantage in the multitrillion-dollar industry that will eventually be off-world living.

A Toy Story

Perhaps you're thinking Made in Space is more the entrepreneurial exception than the rule. After all, Kemmer, Dunn, and Chen might not have known much about 3-D printing, but they were already students at Singularity University, giving them both access to the technology (there are 3-D printers on site) and exposure to all these exponential ideas. But that wasn't the case with Alice Taylor, a British designer who had none of these advantages yet has already made considerable progress toward disrupting the $3.5 billion doll segment of the $34 billion toy industry.[20]

Taylor spent her career in digital media, first as a website creator, later on the digital side of the BBC, and finally as the commissioning editor for education at Channel 4 in London, where much of her job was to make award-winning educational video games.[21] Her interest in games led to an interest in toys, which led her to the doll industry—another linear business ripe for disruption.

Over the past thirty years, the toy business has been transformed. A once-domestic enterprise populated by individual artisans has morphed into a handful of large corporations using overseas mass manufacturers. To be competitive, dolls need to be made in bulk, using an injection mold process that requires one mold for each doll part. Given that each mold can cost tens of thousands of dollars to create, the start-up costs for a single doll can run you hundreds of thousands of dollars.

But maybe not.

Taylor is married to the science fiction writer Cory Doctorow, who knew a little about 3-D printing. (Doctorow, in a sad bit of prophecy, wrote a 2009 book, *Makers*, about how 3-D printers were being used by criminals and terrorists to make AK-47s.)[22] She decided to see if 3-D printing offered an alternative to the traditional—that is, expensive and mass produced—making of dolls. In essence, Taylor set out to see if Roger's third industrial revolution could be applied to toys, as well as cars and rockets.

"The problem," explains Taylor, "was I didn't know much of anything about 3-D printing. So I went to the forum section of Shapeways.com (a 3-D printing marketplace) and found a guy who had posted: 'I can 3-D model for 3-D printing. Hire me.' So I did." Taylor emailed her doll sketches and got a 3-D model back, then printed a real doll from the file. "It was eighteen centimeters high, had no eyes, no hair, and cost me two hundred and twenty pounds, but it existed. It was magical—I just made a doll. I'd never made a doll in my life. I had the same feeling of awe and potential that I had in the early days of the Internet. So I went home and quit my job and set out to build MakieLabs—a company that would allow anyone to custom-design and print a doll."

These days, MakieLabs is entirely powered by 3-D printers. "In our offices we have three small MakerBot printers for prototyping," explains Taylor. "Once the design is right, we then print the final product using large 3D Systems printers on the cloud. We avoid both the huge capital expense required by tooling and, by using on-demand cloud printing, we don't need to buy the large production 3-D printers ourselves. All of our packing, shipping, and marketing can now be virtualized. We don't have warehouse costs, don't need to travel back and forth to the Far East. We don't even need to print our packaging in large batches. We print them as we need them."

Taylor also sees dolls as only the beginning. "Any industry where the end product can be customized is vulnerable," she says. "A doll is just a 3-D shape. But so are a dinosaur, a robot, and a car. We're moving to a world of one-stop manufacturing. We'll either have these tools in our homes and offices or we'll rent them via the cloud. We're at the front end of a very creative time—a great time for disruptive entrepreneurs."

As a way of closing out this chapter, and to provide you with a clearer view of what other industries are immediately ripe for exponential disruption via 3-D printing, take a look at the chart on page 28. It's an analysis by Deloitte Consulting that highlights several of the areas currently experiencing the heaviest 3-D impact and thus ripe with the most entrepreneurial possibility.

CHAPTER THREE

Five to Change the World

The Exponential Landscape

In the last chapter, we took a closer look at exponential growth and entrepreneurial possibility through the lens of additive manufacturing. Yet 3-D printing is only one of the many powerful exponential technologies now moving from deception to disruption. In this chapter, we'll overview five more technologies also ripe for entrepreneurial exploitation: networks and sensors, infinite computing, artificial intelligence, robotics, and synthetic biology. Our aim is to highlight the fundamentals: Where this technology is today, where it will be in a few years from now, and where the hidden opportunities are—areas currently off the radar yet poised for explosion over the next three to five years.

Networks and Sensors

A network is any interconnection of signals and information—the human brain and the Internet being the two most prominent examples. A sensor is a device that detects information—temperature, vibration, radiation, etc.—and when hooked up to the network, can also transmit that information. Right now, both sectors are exploding.

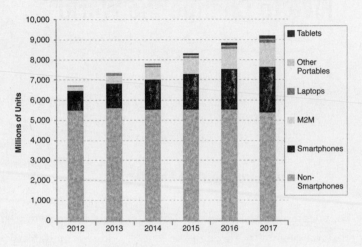

Global Mobile Devices and Connections

Source: https://www.mauldineconomics.com/bulls-eye/

There are over seven billion smartphones and tablets in existence. Each of these devices is a mix of sensors—pressure-sensitive touch screens, microphones, accelerometers, magnetometers, gyros, cameras—that are increasing in number with every new generation of technology. Consider capacitive touch screens—like those found in iPads and iPhones. In 2012, the total area covered by these sensors was 12 million square meters, or enough to blanket two thousand football fields. By 2015, that number will balloon to 35.9 million square meters, or enough to overlay half of Manhattan.[1]

And it's not just communication devices. A similar pattern is playing out in all our "things," transforming a world that was once passive and dumb into one that is active and smart. Take the transportation sector. Today there are sensors in our cars to help us navigate, in our roads to help us avoid traffic jams, and in our parking lots to help us find open spaces. Commercial aircraft are also in the mix. General Electric—which manufactures and leases jet engines to all major airlines—now puts up to 250 sensors in each of their 5,000 leased engines,[2] allowing their health to be monitored in real time, even in midflight. And if the readings fall outside of prescribed levels, GE can swoop in and do a preemptive fix.

Security-related sensors have also exploded onto the scene. Today's all-pervasive video surveillance cameras, now coupled to databases stocked with 120 million facial images, give law enforcement unprecedented search capability. But beyond looking for trouble, our sensors can listen as well. Take ShotSpotter,[3] a gunfire detection technology that gathers data from a network of acoustic sensors placed throughout a city, filters the data through an algorithm to isolate the sound of gunfire, triangulates the location within about ten feet, then reports it directly to the police. The system is generally more accurate and more reliable than information gleaned from 911 callers.

While transportation and security are sectors primarily dominated by larger companies, this doesn't mean that entrepreneurs have not taken advantage of these same exponential trends. As a 2012 *Wired* article pointed out:[4] "Hackers [have begun] using increasingly inexpensive sensors and open source hardware—like the Arduino controller—to add intelligence to ordinary objects." There are now kits that let your plants tweet when they need to be watered, Wi-Fi-connected cow collars that let farmers know when their animals are in heat, and a beer mug that can tell you how much you've drunk during Oktoberfest. As Arduino hacker Charalampos Doukas says, as sensor prices crash downward, "The only limit is your imagination."

To look at this from a more expansive angle, consider that we now live in a world where Google's autonomous car can cruise our streets safely because of a rooftop sensor called LIDAR—a laser-based sensing device that uses sixty-four eye-safe lasers to scan 360 degrees while concurrently generating 750 megabytes of image data per second to help with navigation.[5] Pretty soon, though, we'll live in a world with, say, two million autonomous cars on our roads (not much of a stretch, as that's less than one percent of cars currently registered in the United States),[6] seeing and recording nearly everything they encounter, giving us near-perfect knowledge of the environment they observe. What's more, ubiquitous imaging doesn't stop there.

360-degree LIDAR imaging in Google's driverless car

Source: http://people.bath.ac.uk/as2152/cars/lidar.jpg

In addition to these autonomous cars scanning the roadside, by 2020, an estimated five privately owned low-Earth-orbiting satellite constellations will be imaging every square meter of the Earth's surface in resolutions ranging from 0.5 to 2 meters.[7] Simultaneously, we're also about to see an explosion of AI-operated microdrones buzzing around our cities and taking images down in the centimeter range. Do you want to know how many cars are in your competitor's parking lot in Moscow or Mumbai? Or how about following your competition's supply chain as trucks or trains deliver raw materials to their plant and final product to their warehouses? No problem.

All told, according to a report released by the 2013 Stanford University TSensors Summit, the number of sensors in the world is expected to grow into the "trillions" by 2023.[8] And this is merely the sensor side of the equation.

Trillion Sensor Visions

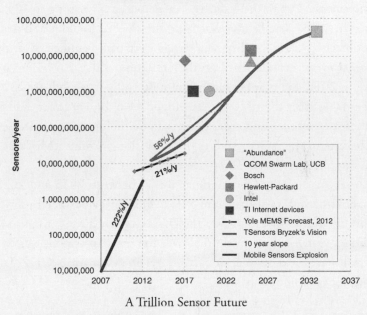

A Trillion Sensor Future

Source: www.futuristspeaker.com/wp-content/uploads/Trillion-Sensor-Roadmap.jpg

Both in speed and in the number of connected devices, networks are undergoing a similar explosion. On the speed side, consider that in 1991, early 2G networks clocked in at a hundred kilobits per second. A decade later, 3G networks hit one megabit per second, while today's 4G networks sport up to eight megabits per second.[9] But in February 2014, Sprint announced plans for Sprint Spark, a super-high-speed network able to deliver 50 to 60 megabits per second to your mobile phone, with a vision of tripling that over time.[10] "Our goal is to support a new generation of online gaming, virtual reality, advanced cloud services, and other applications requiring very high bandwidth," said Stephen Bye, Sprint's CTO. "In more concrete terms, when deployed, Spark will allow you to download a twenty-megabyte video game in three seconds and a one-hour-long high-definition movie in under two and a half minutes." And Sprint is already making progress, having demonstrated an over-the-air speed of 1 gigabit per second at their Silicon Valley lab.

On the connection front, ten years ago, the world had 500 million devices hooked up to the Internet. Today that number is up to 12 billion. "In 2013," says Padma Warrior,[11] the chief technology and strategy officer of Cisco, "eighty new things were being connected to the Internet every second. That's nearly 7 million per day, 2.5 billion per year. In 2014, the number reached almost 100 per second. By 2020, it'll grow to more than 250 per second, or 7.8 billion per year. Add all of these numbers up and that's more than 50 billion things connected to the Internet by 2020." And it's this explosion of connectivity that is building the Internet-of-Things (IoT).

A recent study by Cisco estimated that between 2013 and 2020, this uber-network will generate $19 trillion in value (net profit).[12] Think about this for a moment. The U.S. economy hovers around $15 trillion a year. Cisco is saying that over the ten-year period, this new net will have an economic impact greater than America's GDP. Talk about the land of opportunity.

Global Internet Device Installed Base Forecast

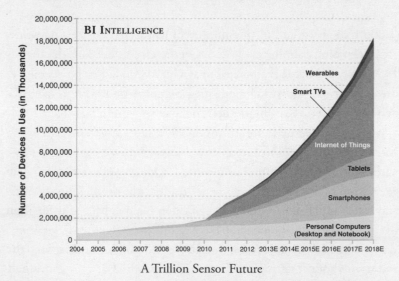

A Trillion Sensor Future

Source: http://www.businessinsider.com/decoding-smartphone-industry-jargon-2013-11

"E" refers to "Estimated", as in estimated size of the market.

So where does that opportunity lie exactly? Well, most researchers feel that there are two critical categories worth exploring: information and automation. Let's start with the former.

Our world of networks and sensors generates enormous quantities of information, much of which is extremely valuable. Take traffic data. A decade back, Navteq built a network of in-road sensors across 400,000 kilometers of Europe (through thirty-five major cities and thirteen European countries).[13] In October 2007, mobile phone giant Nokia (now owned by Microsoft) paid $8.1 billion for that network.[14] Fast-forward five years to mid-2013, when Google paid $1 billion to acquire Waze, an Israeli-based company that generates maps and traffic information, not via electronic sensors, but instead via crowd-sourced user reports—i.e., human sensors, generating maps by using GPS to track the movements of some 50 million users, then generating traffic-flow data as those users voluntarily share information about slowdowns, speed traps, and road closures in real time.[15]

Behavior tracking is another fast-growing information category.[16] Insurance companies putting sensors in cars and pricing policies according to real-time driving behavior is one example. Another is Turnstyle Solutions,[17] a Toronto-based start-up that uses Wi-Fi transmission from smartphones to follow customers around stores, gathering data on where they linger as they shop. Behavior tracking for health care is also growing. AdhereTech[18] now makes smart pill bottles with sensors embedded in them to better ensure patient compliance, while CoheroHealth has combined sensor-enabled inhalers and mobile apps[19] so kids with chronic asthma can track and control their symptoms. These medical applications will keep coming. According to William Briggs, Chief Technology Officer for Deloitte Consulting,[20] "the value of the IoT-related healthcare sector will be a multi-trillion dollar market within the next one to two decades."

Turning our attention to automation—which is essentially the process of gathering all the data collected by the IoT, turning it into a series of next actions, and then, without human intervention, executing those actions. Already, we've seen the first wave of this in the smart assembly lines and supply chains (what's technically called process optimization) that have enabled things like just-in-time delivery. With the smart grid for energy and the smart grid for water—what's technically called resource consumption optimization—we're seeing the second wave. Next up is the automation and control of far more complex autonomous systems—such as self-driving cars.

There are even further opportunities in finding simpler ways to connect decision makers to sensor data in real time. The aforementioned plants that tweet their owners when they need watering were an early (2010) iteration of this sector. A more contemporary example (2013) is the Washington, DC-based start-up SmartThings, a company that CNN called "a digital maestro for every object in the home."[21] SmartThings makes an interface that can recognize over a thousand smart household objects, from temperature sensors that control the thermostat to door and windows sensors that tell you if you left something unlocked to ways to have appliances automatically shut off before you go to bed.

Of course, any discussion of networks and sensors leads directly to a discussion about how we're going to extract value from all this data. The answer is where we're going next. Welcome to the radical world of infinite computing.

Infinite Computing: The Beauty of Brute Force

In late August 2013, Carl Bass, CEO of the software and design giant Autodesk, gave me a tour of his newly constructed Pier 9 center, located at the tail end of San Francisco's Embarcadero.[22] A self-described big kid from Brooklyn (he's 6 feet 5), Bass is clad in jeans, a T-shirt, and a baseball cap. His facility, meanwhile, is clad in the very latest in 3-D printing equipment, machine shop tools, design stations, laser cutters, and welding machines. It's a maker's paradise. These are the tools that turn imagination into reality, and they're all guided by Autodesk's design software, which in turn is powered by infinite computing.

Infinite computing is the term Bass uses to describe the ongoing progression of computing from a scarce and expensive resource toward one that is plentiful and free. Just three or four decades ago, if you wanted to access a thousand core processors, you'd need to be the chairman of MIT's computer science department or the secretary of the US Defense Department. Today the average chip in your cell phone can perform about a billion calculations per second.

Yet today has nothing on tomorrow. "By 2020, a chip with today's processing power will cost about a penny," CUNY theoretical physicist Michio Kaku explained in a recent article for *Big Think*,[23] "which is the cost of scrap paper. . . . Children are going to look back and wonder how we could have possibly lived in such a meager world, much as when we think about how our own parents lacked the luxuries—cell phone, Internet—that we all seem to take for granted."

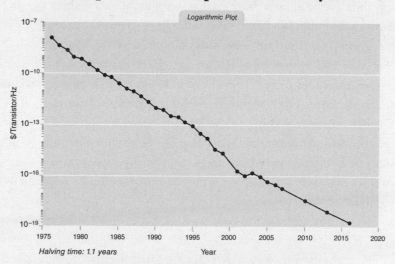

Microprocessor Cost per Transistor Cycle

Source: *www.singularity.com/images/charts/MicroProcessCostPerTrans.jpg*

It's for this reason that Bass feels that much of our thinking about computing is completely backwards. "We've been treating computing as this precious resource," he says, "when really it's abundant. If you look at all the trends, computing is decreasing in cost, increasingly available, increasingly powerful, and increasingly elastic. Every year we produce more computing power than the sum of all prior years. This overabundance is the beginning of a new era."

In the old era, the world of human creation, the so-called designed world was the product of "in the box" thinking—thinking limited by computing scarcity. "In that era," explains Bass, "a problem that used to take one CPU 10,000 seconds to solve would cost about twenty-five cents. But in the new exponential era, powered by near infinite computing, we can now simultaneously apply 10,000 CPUs to the same problem and solve it in one second. Solving the problem 10,000 times faster still costs twenty-five cents, but for the first time in history we're able to apply infinitely more resources to a problem for no additional cost."

Companies like Google, Amazon, and Rackspace are facilitating this shift. Each has assembled massive computational facilities, opened them to the public, and called them the cloud. "Before the cloud, starting a tech company was slow and painful," says Graham Weston,[24] chairman and cofounder of Rackspace. "First you order servers from a vendor like Dell or HP. They show up weeks later. Then you have to configure them, install them, purchase software, load software, then finally connect everything to the Internet. All this took weeks, if not months, and required staff. Today you can go to a provider like Rackspace and minutes later have access to as many servers as you need. And it's massively scalable—vertically or horizontally. It's on-demand capability."

Yet the goal here isn't to become Rackspace (or Amazon or Microsoft), it's to build your big idea atop their infrastructure. Entrepreneurs no longer have to lay out scarce cash for expensive equipment, spend months to install, configure, and program that equipment, worry about what happens when they need to scale up that equipment, or worry about what happens when it breaks or becomes obsolete.

"The cloud is democratizing our ability to leverage computing on a massive scale," says Weston. "Today the computation speed that someone in the middle of Mumbai has access to outstrips what the entire US government had during the sixties and seventies. We're entering an epic period of global innovation where high-performance computing is abundant, reliable, and affordable."

So what does all of this computing power buy you? An entirely new way to approach innovation, for starters. Consider brute force, a term that refers to our newfound ability to use infinite computing to literally surround problems. Imagine you wanted to solve a Sudoku puzzle. You could try and build an elegant mathematical approach—derive an algorithm that calculates the correct missing numbers—or you could simply ask a computer to try every possible number in every possible box and then select the one that works best. The latter approach is brute force.

On my tour of Autodesk's Pier 9 Design Center, as a way of illustrating brute force further, Bass points out an electric go-kart he's building with his fifteen-year-old son. "In the old days, when it comes to attaching this electric motor to that go-kart, I would try for an elegant solution—taking an educated guess on the thickness of brackets and best location, then run a few calculations to find out if what I was doing was adequate. Today I can create a computer model and know exactly the stress and strains at every location for my chosen design. But in the near future, with infinite computing, I could ask the cloud to run design simulations, experimenting with every possible location for the motor and a range of different materials and thicknesses, resulting in not just an adequate design, but the best design."

And what Bass can do, you can do. If your passion is building better go-karts, the technology now allows you to build the best possible go-karts—and in a fraction of the typical time and for a fraction of the typical cost.

And what is true for go-karts is also true for anything else one wants to create. Moreover, we all learn from our mistakes, but until recently, mistakes were too costly for entrepreneurs to make with wanton abandon. This too has changed. Infinite computing demonetizes error-making, thus democratizing experimentation. No longer do we have to immediately dismiss outlandish ideas for the waste of time and resources they invariably incur. Today we can try them all.

Infinite computing has led to a massive increase in design possibilities, though to really unleash this power, you still have to gather the data, feed it into the computer, then code the algorithms to analyze the data. But what if you didn't? What if you could just talk to your computer and your computer could perfectly understand your desires, gather the data for you, and analyze it in a fashion that would answer your question? Well, in the broadest sense, that's the capability we'll explore in our next exponential technology, the exploding field of artificial intelligence.

Artificial Intelligence (AI): Expertise on Demand

"Just what do you think you're doing, Dave?"

Strange fact: These words—their deadly calm and deep menace—were the absolute height of artificial intelligence for nearly fifty years. The line is spoken by HAL, the sentient computer onboard the spaceship *Discovery One*, in director Stanley Kubrick's legendary *2001: A Space Odyssey,* which he cowrote with Arthur C. Clarke.[25] When not fussing about Dave, HAL supports the crew, acting as their interface to the ship, answering questions and helping analyze collected data (the ship is on a scientific mission). HAL's physical presence is depicted as an ominous red television camera eye located on equipment panels throughout the ship.

But that glowing red eye was so last century.

Move over, HAL, say hello to JARVIS.[26] Short for Just Another Rather Very Intelligent System, JARVIS first appeared in *Iron Man* as

Tony Stark's personal AI. Programmed to speak with a male voice in a British accent, JARVIS handles everything from house security to Iron Man suit fabrication to running Stark's global multibillion-dollar business conglomerate—an enormous workload for an extraordinary system.

From a technological perspective, what makes JARVIS extraordinary is both its pervasiveness in Stark's life and its ability to understand natural-language instructions, even when the banter is laden with irony or humor. More technically, JARVIS is a software shell that interfaces between Stark's every desire and the rest of the world, able both to gather data from billions of sensors and to take action through any system or robotic device connected to the AI. In this way, the Internet of Things serves as JARVIS's eyes, ears, arms, and legs.

For sure, JARVIS has dethroned HAL, now holding the title for most recognizable AI in the world, but what makes his dominance more spectacular is that unlike the never-actualized HAL, key elements of JARVIS are starting to come into existence in laboratories and companies around the world.

AI expert and Singularity University cofounder/chancellor Ray Kurzweil[27] explains: "In the 1960s, when Arthur C. Clarke conceived of HAL," explains Kurzweil, "it was clearly science fiction. Fifty years ago, we knew very little about AI. Today it's a different story. Many aspects of JARVIS are either already in existence or on the drawing board."

Kurzweil would know. Bill Gates called him "the best person I know at predicting the future of artificial intelligence." Larry Page hired him as Google's director of engineering, where Kurzweil is leading efforts to develop an AI with natural-language understanding, meaning he's teaching computers to understand the subtle nuances of the spoken and written language, allowing us to ask our machines far more complex questions than "Siri, where can I find a cup of coffee?"

"It's a shift from computers having only logical intelligence to ones that also have emotional intelligence," says Kurzweil. "Once that occurs, AIs will be funny, get your jokes, be sexy, be loving, and even be creative."

Along these lines, in March 2013, I stood on stage at TED, along-side TED curator Chris Anderson, and announced our intent to join forces and design an AI XPRIZE.[28] "Here's the concept," said Anderson. "An XPRIZE for TED to be awarded to the first artificial intelligence to appear on this stage and present a TED talk so compelling that it commands a standing ovation from you the audience."

This concept demands that a key number of AI's abilities either equal or surpass human abilities. When this will happen has been a famous and longstanding debate. Kurzweil himself has pegged the date when AIs will do everything better than humans at 2029.[29] (As explained in *Abundance,* his predictions are based on exponential growth curves and have an amazing track record for accuracy.) Certainly, for most entrepreneurs, 2029 is a date too far out to serve as a basis for a business. But no need to wait, as AI is yet another technology transitioning from deceptive to disruptive, and about to become ubiquitous in our daily lives. Consider, in fact, our daily lives. Today, in America, 80 percent of jobs revolve around the service industry,[30] which in turn can be broken down into four fundamental skills: looking, reading, writing, and integrating knowledge. How far has AI progressed? Computers can now perform all four of these skills and in many cases, better than humans.

Let's take a closer look.

All four of these skills have emerged from a branch of AI known as machine learning—which is literally the science of how machines learn. And one thing for certain, today machines are learning faster than ever.

Looking, the first category, has long been a task better performed by humans than computers. "The first time that a machine learning algorithm was able to 'see' at a level of accuracy similar to humans was in 1995," explains Singularity University's head of machine learning, Jeremy Howard.[31] "That year a US Postal Service competition was won by an algorithm called LeNet 5, which was able to recognize numbers in a zip code and help sort the mail."

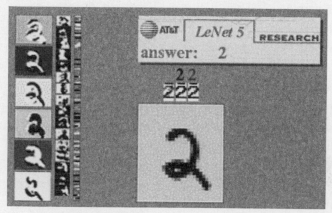

LeNet 5 algorithm recognizing a handwritten "2"

Source: http://yann.lecun.com/exdb/lenet/

Progress remained steady (but unremarkable) until 2011, when a series of major breakthroughs put the machine-learning world on high alert. In Germany, an annual competition pits humans against machine learning algorithms in an attempt to see, identify, and categorize traffic signs. Fifty thousand different traffic signs are used—signs obscured by long distances, by trees, by the glare of sunlight. In 2011, for the first time, a machine-learning algorithm bested its makers, achieving a 0.5 percent error rate, compared to 1.2 percent for humans.[32]

Even more impressive were the results of the 2012 ImageNet Competition, which challenged algorithms to look at one million different images—ranging from birds to kitchenware to people on motor scooters—and correctly slot them into a thousand unique categories. Seriously, it's one thing for a computer to recognize known objects (zip codes, traffic signs), but categorizing thousands of random objects is an ability that is downright human. Only better. For again the algorithms outperformed people.[33]

Similar progress is showing up in reading. Today, there are AIs that can accurately and consistently decipher everything from high school student essays to complicated tax forms far faster than humans. Take legal documents, a linguistic quagmire if ever there was one. Yet, as John Markoff wrote in a 2011 article for the *New York Times*:[34] "Thanks to advances in artificial intelligence, 'e-discovery' software can analyze documents in a fraction of the time for a fraction of the cost. . . . Some programs go beyond just finding documents with relevant terms at computer speeds. They can extract relevant concepts—like documents relevant to social protest in the Middle East—even in the absence of specific terms, and deduce patterns of behavior that would have eluded lawyers examining millions of documents."

In our third human-skill category—writing—a January 2014 Deloitte University Press report[35] explains that AI is making a dent here too. "Intelligent automation, though still rapidly developing, has matured to the point where it has penetrated nearly every sector of the economy. [In the writing category], Credit Suisse uses a technology from Narrative Science to analyze millions of data points on thousands of companies and automatically write English research reports that assess company expectations, upside, and risk. The reports help analysts, bankers, and investors make long-term investment decisions and has tripled the volume of reports produced while improving their quality and consistency compared with analyst-written reports."

Integrating knowledge, our fourth skill, represents the much more complex ability to pull together information from many sources and reach accurate conclusions. Here we find arguably the most important breakthrough and the greatest entrepreneurial opportunity. Remember IBM's Watson, the supercomputer who bested humans on *Jeopardy*[36] in February 2011? Well, as of November 2013, IBM has uploaded Watson to the cloud, making it a development platform available to anyone, especially entrepreneurs. As Michael Rhodin,[37] the senior vice president at IBM in charge of Watson, says, "Putting Watson on the cloud is aimed at spurring innovation and fueling a new ecosystem of entrepreneurial application software providers—ranging from start-ups and emerging venture-capital-backed business to established players. We've even established a new $100 million venture fund to back start-ups using Watson."

One example of a new start-up backed by Watson is Modernizing Medicine. Back in 2011, Modernizing Medicine launched as an iPad-based, specialty-specific electronic medical records platform with a cool crowdsourced twist.[38] For example, all dermatologists who sign up with Modernizing Medicine have their outcome data—that is, what was wrong with a patient and what treatment they prescribed— de-identified (meaning patients' names are removed) and aggregated. This information then becomes available to every dermatologist on the network—some 3,000 of them, or 25 percent of all dermatologists in America—thus significantly improving quality of care. But, in 2014, Modernizing took a huge step forward and partnered with Watson. Since winning on *Jeopardy*, Watson has been sent to medical school—loaded up with millions of journal articles, textbooks, patient outcomes, scientific papers, and the like. By combining their structured patient outcome data with Watson's unstructured research data, doctors on the Modernizing Medicine network have access to incredible levels of point-of-care information. "It would be impossible for us humans to replicate what Watson does in health care," says Modernizing Medicine CEO Daniel Cane.[39] "Not only can it answer questions pulled from millions of individual documents, it can instantly cite the source and confidence level. Beyond empowering physicians with the most powerful Q&A tool ever created, it will fundamentally change the practice of medicine."

Even better, for entrepreneurs interested in building Watson-backed business, Cane was stunned by how easy it was to work with IBM. "They provided so much support and guidance," he explains, "that we were able to build our entire Watson-powered prototype in two weeks."

One of this book's core goals is to point out those pivotal moments when a technology becomes ready for entrepreneurial prime time. Watson in the cloud, tied to an openly available API, is the beginning of one such moment, the potential for a Mosaic-like interface explosion, opening AI to all sorts of new businesses and heralding its transition from deceptive to disruptive growth. Attention, exponential entrepreneurs: What are you waiting for?

And everything we've just covered is here today. "Soon," says Ray Kurzweil,[40] "we will give an AI permission to listen to every phone conversation you have. Permission to read your emails and blogs, eavesdrop on your meetings, review your genome scan, watch what you eat and how much you exercise, even tap into your Google Glass feed. And by doing all this, your personal AI will be able to provide you with information even before you know you need it."

Imagine, for example, a system that recognizes the faces of people in your visual field and provides you with their names. This shouldn't be too much of a mental stretch, as these capabilities are already coming online. Now imagine that this same AI also has contextual understanding—meaning the system recognizes that your conversation with your friend is heading in the direction of family life—so the AI reminds you of the names of each of your friend's family members, as well as any upcoming birthdays they might have.

Behind many of the AI successes mentioned in this section is an algorithm called Deep Learning. Developed by University of Toronto's Geoffrey Hinton for image recognition, Deep Learning has become the dominant approach in the field. And it should come as no surprise that in spring of 2013, Hinton was recruited, like Kurzweil, to join Google[41]—a development that will most likely lead to even faster progress.

More recently, Google and NASA Ames Research Center—one of NASA's field centers—jointly acquired a 512 qubit (quantum bit) computer manufactured by D-Wave Systems to study machine learning. With lightning speed, this computer can tackle face and voice recognition, as well as understanding biological behavior and the management of very large systems. "The tougher, more complex the problem," says Geordie Rose,[42] D-Wave's cofounder and CTO, "the better the results. For most problems, it was eleven thousand times faster, but in the 'more difficult' category it was thirty-three thousand times faster. In the 'most difficult' category, it was fifty thousand times faster." So when Stark asks JARVIS to look at a massive amount of imagery data and pick out certain faces in the crowd, well, JARVIS is probably using qubits.

Why am I telling you about artificial intelligence aided by quantum computers? Not because I expect you to start developing these machines or using quantum computing (though a new SU start-up called 1Qbit[43] has created an online user-interface that would allow an entrepreneur to get access to a D-Wave machine via the web). Instead, the point is that AI has been in a deceptive phase for the past fifty years, ever since 1956, when a bunch of top brains came together for the first time at the Dartmouth Summer Research Project[44] and made a "spectacularly wrong prediction" about their ability to crack AI over a single hot New England summer. But today, couple the successes of Deep Learning and IBM's Watson to the near-term predictions of technology oracles like Ray Kurzweil, and we find a field reaching the knee of the exponential growth curve—that is, a field ready to run wild in disruption.

So what does this mean to you, the exponential entrepreneur? This is a multibillion-dollar question. But as you try to find answers, remember that JARVIS is essentially the ultimate user interface, democratizing every exponential technology and giving all of us access to Stark-like capabilities.

Robotics: Our New Workforce

Camel racing is a centuries-old tradition in the Middle East, but it's an activity primarily reserved for large festivals. Yet, in the past half century, the sport has been transformed into both a mainstay of Arab culture— think the Kentucky Derby for sheikhs—and one of the richest sports on Earth. It's the jockeys who have changed the most. Twenty years ago, camels were ridden by children—the lightest possible riders—but general principle, injury, and death led to a humanitarian outcry. So both the UAE and Qatar banned the practice, instead replacing children with an even lighter saddle occupant—the robot jockey.[45]

Today, in camel racing, robot jockeys are the norm. Exactly like traditional jockeys, these robo-replacements sit on a saddle, steer with the reins, and prod with a whip. To prevent the camels from being frightened by their cyborg occupants, designers found that human-like features—a mannequin face, sunglasses, a hat, traditional racing silks, and even the traditional perfumes used by human jockeys—help keep the animals calm. The latest robot jockeys are small, about a foot high, and light, weighing between five and eight pounds, with skinny hinged arms that control the reins and whip. There's even a speaker on the robot so camel owners can issue commands to their animals via walkie-talkie as they follow along on an outside track (in air-conditioned SUVs).

Of course, our point isn't that there's a bevy of entrepreneurial pos-sibility in camel racing. It's that robotics, another exponential technol-ogy long mired in deception, is now heading for disruption. According to a report by the Littler Workplace Policy Institute:[46] "Robotics is the fastest growing industry in the world, poised to become the *largest* in the next decade." Which is to say, robot jockeys are just the beginning.

Consider Baxter,[47] the brainchild of legendary roboticist Rod-ney A. Brooks, Panasonic Professor of Robotics (emeritus) at MIT and cofounder of iRobot (creator of the Roomba). With a humanoid design, a nine-foot wingspan, and a tablet computer for a face, Baxter looks like something out of a cartoon. Grab one of his arms, for exam-ple, and Baxter will turn his head in your direction, the tablet com-puter displaying a pair of wide-open eyes to demonstrate interest. But what is most exciting about Baxter is his user interface.

Unlike most industrial robots, Baxter is human-safe. Getting in a room with a typical six-axis car-building robot is a good way to get dead—which explains why most industrial robots are cordoned off from humans. But Baxter doesn't need a cage. Sensors detect when the robot hits something unexpected and stops the motion immediately, so "he" can't hurt you.

Moreover, Baxter has an elegant and simple user interface. Instead of a complicated code-based programming, it learns through guided imitation. Simply move the robot's arms through the motions you want him to replicate, and presto, he's programmed. And with AI soon coming online, it won't be long before putting Baxter through his motions will be replaced by simply having a conversation with him. "Hey, Baxter, could you put this tire on that car?"

"Baxter is a big step forward," says Dr. Dan Barry,[48] head of robotics at SU. "It's the first robot that bridges the gap between mindless, repetitive, robust, single-purpose industrial robots and intelligent, widely sensing, situationally aware, computationally complex, delicate research robots." More important, Baxter is the kind of robot that entrepreneurs can now build businesses around. Case in point: Digital Apparel, a Bay Area clothing start-up, plans to do 3-D scans of their customer's bodies, then use those scans as a pattern for cutting and stitching denim to make perfect custom-fitting jeans. And what robot will Digital Apparel use to help assemble your jeans? You guessed it, Baxter.

Besides user-friendly robotic interfaces, we're also seeing exponential progress in robotic agility and mobility. Enabled by a new generation of sensors and actuators, and driven by near-infinite computing and artificial intelligence, there's a Cambrian explosion[49] in robotics, with species of all sizes, shapes, and modes of mobility crawling out of the muck of the lab and onto the terra firma of the marketplace. Festo, for one example, has created a robot that flies like a bird. Boston Dynamics, for another, now makes robots that can climb, crawl, jump, and hop, and all while carrying heavy loads (some bots can manage over a hundred kilograms of weight). These "Sherpa-bots" can traverse boulder-strewn hillsides, balance on sheets of ice, and even jump from the ground to a rooftop three stories up.

But what has been relatively slow progress—run out of university labs and funded by government grants—took a quantum leap forward in late 2013, when Amazon announced it was going into the drone business[50] and Google announced the acquisition of eight robotics companies (including Boston Dynamics).[51] With the big dogs in the game, progress is coming even faster.

And the resulting change will be considerable. Robots don't unionize, don't show up late, and don't take lunch, yet Baxter can work an assembly line for the equivalent of $4 an hour.[52] A 2013 report from the Oxford Martin School concludes that 45 percent of American jobs are at high risk of being taken by computers (AI and robots) within the next two decades.[53] Good or bad, this same trend is evident around the world. In China, Foxconn, the Chinese electronics manufacturer that builds Apple's iPhone, made news in 2013 when the skyrocketing demand for cell phones led to labor disputes, reports of harsh working conditions, even riots and suicides. In the aftermath of these reports, Foxconn's president, Terry Gou, said he intended to replace one million workers with robots over the next three years.[54]

Besides replacing our blue-collar workforce, over the next three to five years, robots will invade a much wider assortment of fields. "Already," says Dan Barry, "we're seeing telepresence robots transport our eyes, ears, arms, and legs to conferences and meetings. Autonomous cars, which are, after all, just robots, will [start to] chauffeur people around and deliver goods and services. Over the next decade, robots will also move into health care, replacing doctors for routine surgeries and supplementing nurses for eldercare. If I were an exponential entrepreneur looking to create tremendous value, I'd look for those jobs that are least enjoyable for humans to do. . . . Given that the global market for unskilled labor is worth many trillions of dollars, I would say this is a huge opportunity."

So how does an entrepreneur take advantage of this opportunity? Well, as a June 3, 2013, article in *Entrepreneur* explained:[55] "Startup infrastructure dedicated to robotics is likewise emerging: hacker spaces (Robot Garden), accelerators (Robot Launchpad) and even a dedicated venture capital firm, New York City–based Grishin Robotics, founded last June by Russian internet entrepreneur Dmitry Grishin." These advances are proof that a field long out of reach for most entrepreneurs is now open for business. "We're seeing very early indicators that this market is coming into fruition immediately," says Jon Callaghan, a founding partner in the early-stage tech venture capital firm True Ventures, in that same *Entrepreneur* article. "It's super early, but it will hit very, very quickly, and we'll look back on 2013 . . . as a year for robotics coming into its own."

Genomics and Synthetic Biology

Throughout the past few chapters, we've been examining exponentials poised to explode over the next three to five years and seeing how these technologies reinforce and empower one another—the rise of cloud computing enables more capable and ubiquitous AI, which in turn allows the average entrepreneur to program robots. To close this chapter, we're going to examine synthetic biology, a technology that's a little further out—say, five to ten years—but is still transitioning from deception to disruption.

And it's going to be a sizable disruption.

Synthetic biology[56] is built around the idea that DNA is essentially software—nothing more than a four-letter code arranged in a specific order. Much like with computers, the code drives the machine. In biology, the order of the code governs the cell's manufacturing processes, instructing it to make specific proteins and such. But, as with all software, DNA can be reprogrammed. Nature's original code can be swapped out for new, human-written code. We can co-opt the machinery of life, telling it to produce—well, whatever we can think of.

In itself, this idea isn't new. With genetic engineering, we've been inserting a gene or two from this organism into that organism—such as taking the DNA that makes jellyfish glow and, as South Korean researchers did in 2007, inserting it into cats to create, you guessed it, glow-in-the-dark cats.[57] The difference with synthetic biology is that it's not just individual letters being swapped out—it's whole genomes.

"Synthetic biology is essentially genetic engineering gone digital," explains synthetic biologist and Autodesk distinguished researcher and S. U. professor Andrew Hessel.[58] "Used to be all this stuff was done by hand in the lab, with enormous expense and high error rates. Today we manipulate DNA with computers, using programs that function much like word processors. Mix and match genetic code, spell and error check, shuffle bits around—it's becoming drag-and-drop easy."

This increase in simplicity and accessibility has opened the door to a wonderland of possibilities. New fuels, foods, medicines, construction materials, clothing fibers, and even new organisms are all in the offing. Everything we now manufacture industrially, we'll eventually be able to assemble biologically.

Take your typical day. After you get out of bed in the morning, what's the first thing you do? Brush your teeth. Right now your toothpaste is mostly chalk and flavoring, but with synthetic biology, it can be specifically designed to fight your breed of bad breath microbes. "That's not all," continues Hessel. "It can have tooth-polishing nanoparticles designed to continue cleaning long after you've stopped brushing. It can be designed to detect infection or cancer or diabetes, turning different colors in the presence of each, or to release custom-designed probiotics that balance your microbiome. It can do all of these things. And that's just the first thing you do in the morning."

To many, synthetic biology still sounds like science fiction, but what is transforming it into science fact is the same force driving all the other exponential technologies—Moore's law. Because DNA is nothing more than a four-letter code, when genetics went digital, it was transformed into an information science and thus hopped on the exponential expressway. This is why, in 1995, the National Institute of Health estimated it would take fifty years and $15 billion to sequence the first human genome. But in 2001, Dr. J. Craig Venter completed the task in nine months for $100 million. Today, thanks to exponential growth, you can sequence billions of letters of your genome in a few hours for about $1,000.[59] But here's the kicker: Biotechnology isn't just accelerating at the speed of Moore's law, it's accelerating at five times the speed of Moore's law—doubling in power and halving in price every four months!

What all this means is that bioengineering, once an incredibly exclusive field limited to those with PhDs in large government and university labs, is starting to become an entrepreneurial playground. Already, biohacker spaces (where anyone can go and learn to play with synthetic biology) exist in most major cities and all the necessary equipment is available online (at cut-rate costs). For those not inclined toward the science, dozens of contract research and manufacturing services (CRAMS) are willing to do the heavy lifting for a fee. Perhaps the biggest news is that synthetic biology is on the verge of developing the ultimate enabling technology and leveler of the playing field—a set of user-friendly interfaces.

One such tool is under development at Autodesk's Pier 9 design center, where Carlos Olguin[60] is working on Project Cyborg, a synthetic biology interface that allows high school students, entrepreneurs, and citizen scientists to program DNA. "We're working hard to deskill the technology," says Olguin. "A modeling process that would previously have taken weeks or months to complete and [would] require post-PhD level abilities can now be completed in a few seconds with relative ease. The goal here is to make programming with biological parts as intuitive as Facebook. We want more people designing and contributing, people who don't have a PhD, people like Jack Andraka— the fourteen-year-old high school student who won the grand prize of the Intel Science and Engineering Fair for creating a fast, accurate, pennies-on-the-dollar test for pancreatic cancer."

And because this software lives in the cloud, not only can anyone use it to run experiments, anyone can sell the results on Autodesk's soon-to-be-established Project Cyborg marketplace, meaning synthetic biology is about to get access to that fantastic accelerator of entrepreneurial possibility: its first app store.

Making One Hundred Years Old the New Sixty

While the bulk of this chapter has been concerned with the exploitation of individual technologies for their entrepreneurial possibilities, even more potential can be found at the intersection of multiple fields. In fact, along just these lines, in March of 2013, I joined forces with genetics wizard Dr. J. Craig Venter and stem-cell pioneer Dr. Robert Hariri to found perhaps my boldest venture ever: Human Longevity, Inc. (HLI).[61] Venter, who serves as CEO, described HLI's mission as "using the combined power of genomics, infinite computing, machine learning, and stem cell therapies to tackle one of the greatest medical, scientific, and societal challenges—aging and aging-related diseases." Hariri, who pioneered the use of placental-derived stem cells, goes on to say: "Our goal is to help all of us live a longer and healthier life. By reenergizing our stem cells, the regenerative engine of our bodies, we can maintain our mobility, cognition, and aesthetics long into our later years." Put simpler, HLI's goal is to make one hundred years old the new sixty.

We launched HLI with $85 million in seed capital, raised at record speed. Part of the reason for this velocity is that the company sits at the intersection of many of the exponential technologies discussed in this chapter: robotics, which enables lightning-fast sequencing; AI and machine learning, which can make sense of petabytes of raw genomic data; cloud computing and networks for transmitting, handling, and storing that data; and synthetic biology for correcting and rewriting the corrupted genome of our aging stem cells. Couple that with the incredible value proposition of abundant, longer, and healthier lives—there is over $50 trillion locked up in the bank accounts of people over the age of sixty-five—and you understand the potential.

And understanding this potential is critical if you're going to succeed as an exponential entrepreneur. Consider that, twenty years ago, the idea that a computer algorithm could help companies with funny names (Uber, Airbnb, Quirky) dematerialize twentieth-century businesses would have seemed delusional. Fifteen years ago, if you wanted access to a supercomputer, you still had to buy one (not rent one by the minute on the cloud). Ten years ago, genetic engineering was big government, and big business and 3-D printing meant expensive plastic prototypes. Seven years ago, the only robot most entrepreneurs had access to was a Roomba, and AI meant a talking ATM machine, not a freeway-driving autonomous car. Two years ago, the idea of living past a hundred was a crazy idea. You get the picture.

And it's a radical picture. Today's exponential entrepreneurs have at their disposal more than enough power, as Steve Jobs famously said, to put a dent in the universe.[62] Billion-dollar companies are being built faster than ever, and trillion-dollar industries are on their way. But before you consider taking your swing at the exponential piñata, the first and most important step is to convince yourself that you can take this step—which is why our next three chapters focus on the most critical tools in the kit of an exponential entrepreneur: the psychological techniques needed to go bold.

PART TWO

BOLD MINDSET

CHAPTER FOUR

Climbing Mount Bold

The Secret of "Skonk"

It started sometime in the 1930s. Our location is deep backwoods, Dogpatch, Kentucky, where tragedy is unfolding. Dozens and dozens of locals are being killed on a yearly basis, felled by the toxic fumes of skonk oil, a compound brewed at the so-called Skonk Works by grinding dead skunks and worn shoes inside a blazing still. Or at least, that's how Al Capp tells the story.[1]

Al Capp was the creator of the legendary comic strip *Li'l Abner*, and the Skonk Works were among his most memorable inventions—though Capp had little to do with the term's considerable staying power. Instead, we can thank the aerospace giant Lockheed for that.

In 1943, Lockheed's chief engineer, Clarence "Kelly" Johnson, fielded a call from the US Department of Defense. German jet fighters had just appeared over Europe, and America needed a counterpunch. The mission was unbelievably critical, the deadline impossibly tight. Kelly, though, had an idea. At their Burbank, California, facility, he recruited a small posse of his brightest engineers and mechanics, gave them total design freedom—no idea too weird or too wild—and then walled them off from the rest of Lockheed's bureaucracy. No one with

71

out proper clearance was even told the purpose of this new project. No one with proper clearance ever breathed a word of their mission. Though, as these employees were housed in a rented circus tent (space was at a premium) intentionally located next to an exceptionally stinky plastics factory (to keep nosy people away), breathing itself was a little hard. That was why engineer Irv Culver borrowed Al Capp's terminology and started calling the place the Skonk Works.

One day, the story goes, the Department of the Navy was trying to reach Lockheed for an update on their new jet (technically the P-80) and got transferred to Culver's line by mistake. In his then-standard fashion, he answered the phone: "Skonk Works, inside man, Culver."[2]

The name stuck. A few years later, at the request of the comic strip copyright holders, the spelling was changed to Skunk Works.

And the Skunk Works worked. The US's first military jet was delivered to the Pentagon just 143 days later, a staggeringly short time frame that was, more incredibly, seven days ahead of schedule. In a typical military project, contractors can't even get their paperwork signed in that window, forget about building anything. Yet over the coming decades Lockheed's Skunk Works would repeat this success, going on to produce some of the world's most famous aircraft—the U-2, the SR-71, the Nighthawk, the Raptor—with this same methodology. These planes helped the United States win the cold war, of course, but their bigger impact was organizational: for the next half a century, whenever a company wanted to go bold, skunk was often the way innovation got done.

Everyone from Raytheon and DuPont to Walmart and Nordstrom has gotten in on the skunk game. In the early 1980s, to offer another example, Apple cofounder Steve Jobs leased a building behind the Good Earth restaurant in Silicon Valley, stocked it with twenty brilliant designers, and created his own skunk works to build the first Macintosh computer.[3] The division was set apart from Apple's normal R&D department and led by Jobs himself. When people asked him why they needed this new facility, Jobs liked to say: "It is better to be a pirate than join the Navy."

The question is why. When it comes to fostering bold innovation, why is it better to be a pirate? Why does the skunk methodology consistently foster such great results? And most importantly, what does this have to do with today's entrepreneur and a desire to tackle the bold?

Turns out, plenty.

The Secrets of Skunk: Part One

Over the past few chapters, we've seen how exponentially accelerating technology provides today's entrepreneurs with an astounding reach, allowing small teams of innovators to tackle the kinds of grand challenges that were once the sole province of corporations or governments. This is huge news. At Singularity University, one of our core tenets is that the world's grandest challenges contain the world's biggest business opportunities. And because of exponential technology, for the first time in history, entrepreneurs can actually get in on this game. But there's a rub: Exponentials alone won't get this job done.

Climbing Mount Bold is not just technologically difficult, it's also incredibly psychologically difficult. Every innovator interviewed for this book emphasized the importance of the mental game, arguing that without the right mindset, entrepreneurs have absolutely no chance of success. I couldn't agree more. Attitude is the ball game. If you think you can or think you can't—well, you're right. Thus the goal of part two of this book is to provide you with an attitude upgrade—a series of battle-hardened, time-tested psychological strategies for going big and bold.

Toward this end, we'll take a three-pronged approach. In this chapter, we'll peek under the hood of skunk, getting at the core mechanisms that have turned this approach to innovation into one of the most successful in modern history. In the next, we'll explore the mental tools and techniques that I have personally relied upon in my life and work. Finally, to close part two, we'll meet a group of exceptional entrepreneurial billionaires—Elon Musk, Jeff Bezos, Richard Branson, and Larry Page—who are important not solely because of their financial success, but because that success has given them the ability to think at scale, the very skill needed to tackle grand challenges.

But first, the secrets of skunk.

Traditionally, when exploring these secrets, researchers start by parsing Kelly Johnson's fourteen rules for going skunk. This is a useful approach, and in the next section, we too shall venture down this path. But before that happens, it's helpful to address an idea baked into the DNA of this methodology yet often omitted from the discussion—the purpose of the project itself.

Companies do not go skunk for business as usual. These innovation accelerators are always about business as *un*usual. They are created to tackle the Herculean, purposefully built around what psychologists call "high, hard goals." And it's the difficult nature of those goals that is actually the first secret to skunk success.

In the late 1960s, University of Toronto psychologist Gary Latham and University of Maryland psychologist Edwin Locke discovered that goal setting is one of the easiest ways to increase motivation and enhance performance.[4] Back then this was something of a shocking finding. General thinking was that happy workers were productive workers and putting too much stress on employees—by, say, imposing goals—was considered bad for business. But in dozens and dozens of studies, Latham and Locke found that setting goals increased performance and productivity 11 to 25 percent.[5] That's quite a boost. If an eight-hour day is our baseline, that's like getting two extra hours of work simply by building a mental frame (aka a goal) around the activity.

But not every goal is the same. "We found that if you want the largest increase in motivation and productivity," says Latham, "then big goals lead to the best outcomes. Big goals significantly outperform small goals, medium-sized goals, and vague goals. It comes down to attention and persistence—which are two of the most important factors in determining performance. Big goals help focus attention, and they make us more persistent. The result is we're much more effective when we work, and much more willing to get up and try again when we fail."

This is a critical piece of information for the exponential entrepreneur. Starting any business is hard. Starting a business with the intention of disrupting an industry—now, that's downright terrifying. But Locke and Latham's work shows that there's hidden leverage available. Because the practice focuses attention and increases motivation, by setting big goals, we're actually helping ourselves achieve those big goals.

Yet for these high, hard goals to really work their magic, Locke and Latham found that certain moderators—the word psychologists use to describe "if-then" conditions—need to be in place. One of the most important is commitment. "You have to believe in what you're doing," continues Latham. "Big goals work best when there's an alignment between an individual's values and the desired outcome of the goal. When everything lines up, we're totally committed—meaning we're paying even more attention, are even more resilient, and are way more productive as a result."

This is another key point. When Kelly Johnson created the original skunk works, the goal wasn't to build a new plane in record time—that was just one of many things that happened on the way to the main big goal: saving the world from Nazi peril. This is the kind of big goal everyone can get behind. It's why the engineers agreed to work horrific hours in a foul-smelling circus tent. And most importantly, because this alignment between core values and desired outcomes jacked up performance and productivity, it became one of the fundamental reasons that plane was delivered in record time.

The Secrets of Skunk: Part Two

At the Lockheed skunk works, Kelly Johnson ran a tight ship. He loved efficiency. He had a motto—"be quick, be quiet, and be on time"—and a set of rules.[6] And while we are parsing the deep secrets of skunk, it's to "Kelly's rules" we must now turn.

Wall the skunk works off from the rest of the corporate bureaucracy—that's what you learn if you boil Johnson's rules down to their essence. Out of his fourteen rules, four pertain solely to military projects and can thus be excluded from this discussion. Three are ways to increase rapid iteration (a topic we'll come back to in a moment), but the remaining seven are all ways to enforce isolation. Rule 3, for example: "The number of people with any connection to the project should be restricted in an almost vicious manner." Rule 13 is more of the same: "Access by outsiders to the project and its personnel must be strictly controlled by appropriate security measures." Isolation, then, according to Johnson, is the most important key to success in a skunk works.

The reasoning here is twofold. There's the obvious need for military secrecy, but more important is the fact that isolation stimulates risk taking, encouraging ideas weird and wild and acting as a counterforce to organizational inertia. Organizational inertia is the notion that once any company achieves success, its desire to develop and champion radical new technologies and directions is often tempered by the much stronger desire not to disrupt existing markets and lose their paychecks. Organizational inertia is fear of failure writ large, the reason Kodak didn't recognize the brilliance of the digital camera, IBM initially dismissed the personal computer, and America Online (AOL) is, well, barely online.

But what is true for a corporation is also true for the entrepreneur. Just as the successful skunk works isolates the innovation team from the greater organization, successful entrepreneurs need a buffer between themselves and the rest of society. As Burt Rutan, winner of the Ansari XPRIZE, once taught me: "The day before something is truly a breakthrough, it's a crazy idea." Trying out crazy ideas means bucking expert opinion and taking big risks. It means not being afraid to fail. Because you will fail. The road to bold is paved with failure, and this means having a strategy in place to handle risk and learn from mistakes is critical.

In a talk given at re:Invent 2012, Amazon CEO Jeff Bezos[7] explains it like this: "Many people misperceive what good entrepreneurs do.

Good entrepreneurs don't like risk. They seek to reduce risk. Starting a company is already risky . . . [so] you systematically eliminate risk in those early days."

This is exactly where Kelly Johnson's final three rules come into play.

All three of these rules are ways to increase rapid iteration—which is one of the best risk-mitigation strategies ever developed. If you're looking for a quick and dirty definition of the term, try the unofficial motto of Silicon Valley: "Fail early, fail often, fail forward."[8] Bold ventures—especially the world-changing type we're advocating here—require this kind of experimental approach. Yet as most experiments fail, real progress requires trying out tons of ideas, decreasing the lag time between trials, and increasing the knowledge gained from results. This is rapid iteration.

Take software design. The traditional methodology involved creating a product in secret, usually over a number of years, then bomb-dropping it on the public with one massive launch. Unfortunately, in a world of increasingly rapid change, spending a few years separated from one's customers can mean bankruptcy.

Enter agile design, an ideology that emphasizes fast feedback loops.[9] Instead of launching a finely polished gem, companies now release a "minimum viable product," then get immediate feedback from customers, incorporate that feedback into the next iteration, release a slightly upgraded version, and repeat. Instead of design cycles that last years, the agile process takes weeks and produces results directly in line with consumer expectations. This is rapid iteration.

"We saw this with Gmail," says Salim Ismail.[10] "Instead of sending designers off to spend years coming up with the best twenty-five features anyone would ever want in an email program, Google released a version with around three features and asked their customers what else they wanted the program to do. It was very fast feedback and completely iterative. That's why LinkedIn founder Reid Hoffman famously said, 'If you're not embarrassed by the first version of your product, you've launched too late.'"

Motivation 2.0

Up to now, our exploration of skunk has focused on big goals, isolation, and rapid iteration as strategies for tuning psychology and increasing productivity. This section will do more of the same, only instead of examining these strategies in isolation, we'll blend them together, exploring the additional and significant psychological boost that comes from employing these ideas in aggregate.

To understand this boost requires dipping back into the science of motivation. For most of the last century, that science focused on extrinsic rewards—that is, external motivators, "if-then" conditions of the "do this to get that" variety. With extrinsic rewards, we incentivize the behavior we want more of and punish the behavior we dislike. In business, for example, when we want to drive performance, we offer classic extrinsic rewards: bonuses (money) and promotions (money and prestige).

Unfortunately, an ever-growing pile of research shows that extrinsic rewards do not work like most suppose. Take money. When it comes to increasing motivation, cash is king only under very specific conditions. For very basic tasks that don't require any cognitive skill, money can effectively influence behavior. If I'm nailing together boards for five dollars an hour, offering me ten will increase the rate at which I nail. But once tasks become slightly more complex—such as shaping those nailed boards into a house—once they require even the slightest bit of conceptual ability, money actually has the exact opposite effect: It lowers motivation, hinders creativity, and decreases performance.[11]

What's more, this isn't the only issue with money as a motivator. Money, it now appears, is only an effective motivator until our basic biological needs are met, with a little left over for discretionary spending. This is why, in America, as the Nobel laureate Daniel Kahneman recently discovered, when you plot happiness and life satisfaction alongside income, they overlap until $70,000—i.e., the point at which money stops being a major issue—then wildly diverge.[12] Once we pay people enough so that meeting basic needs is no longer a constant cause for concern, extrinsic rewards lose their effectiveness, while intrinsic rewards—meaning internal, emotional satisfactions—become far more critical.

Three in particular stand out: autonomy, mastery, and purpose. Autonomy is the desire to steer our own ship. Mastery is the desire to steer it well. And purpose is the need for the journey to mean something. These three intrinsic rewards are the very motivators that motivate us most. In his book *Drive*,[13] author Daniel Pink explains it like this:

> The science shows that . . . typical twentieth-century carrot-and-stick motivators—things we consider somehow a "natural" part of human enterprise—can sometimes work. But they're effective in only a surprisingly narrow band of circumstances. The science shows that "if-then" rewards . . . are not only ineffective in many situations, but can also crush the high-level, creative, conceptual abilities that are central to current and future economic and social progress. The science shows that the secret to high performance isn't our biological drive (our survival needs) or our reward-and-punishment drive, but our *third drive*—our deep-seated desire to direct our own lives, to extend and expand our abilities, and to fill our life with purpose.

To take on the bold, we need this third drive. Leveraging exponential technology to tackle big goals and using rapid iteration and fast feedback to accelerate progress toward those goals is about innovation at warp speed. But if entrepreneurs can't upgrade their psychology to keep pace with this technology, then they have little chance of winning this race.

And this is another secret to skunk—it provides the full upgrade. Combine the rules for isolation and rapid iteration discussed in this section with the value-aligned big goals from the last and you end up with a great recipe for autonomy, mastery, and purpose. Walling off the innovation team creates an environment where people are free to follow their own curiosity—it amplifies autonomy. Rapid iteration means accelerated learning cycles, which means putting people on the path to mastery. And aligning big goals with individual values creates true purpose.

Most importantly, you don't need to be running a skunk works to take advantage of these intrinsic motivators. Google taps "autonomy" on a company-wide basis with their 20 percent time—engineers are encouraged to devote 20 percent of their time to projects of their own design—and the resulting boost in motivation explains why Googlers often joke that "20 percent time should really be called '120 percent time.'"[14] Tony Hseih, CEO of Zappos, helped disrupt the retail space by emphasizing mastery, making the "pursuit of growth and learning" central to his corporate philosophy and famously saying: "Failure isn't a badge of shame. It is a rite of passage."[15] And Toms Shoes CEO Blake Mycoskie harnessed the power of purpose by deciding to give away one pair of shoes to a child in the developing world for every pair sold.

With this kind of psychological core, it's no surprise that Google, Zappos, and Toms all became industry leaders in record time. Creating a company with autonomy, mastery, and purpose as key values means creating a company built for speed. And this is no longer optional. In a world of increasing rapid change, tapping our third drive is an absolute fundamental for any exponential entrepreneur. Yet, unlike big companies, which often have to go skunk to tap this drive, bold entrepreneurs can get ahead of the game, baking autonomy, mastery, and purpose into their corporate culture from the get-go, rather than bolting them on later.

How Google Goes Skunk

Astro Teller is tall and lean, with a thick goatee, rimless glasses and long hair, usually tied back in a ponytail. He's the grandson of two different Nobel laureates, including Edward Teller, the father of the atomic bomb. He's also the kind of guy who wears T-shirts. Usually the shirts say something pertinent yet ironic. A few years back, for example, when he gave a talk about the importance of innovation to a group of eighty Fortune 200 executives at Singularity University, he was wearing a T-shirt that read: "Safety Third."

Teller heads GoogleX, the Internet giant's skunk works, though his technical title is Captain of Moonshots. The title comes from a conversation he had with CEO Larry Page not long after he was hired. "In the early days," explains Teller,[16] "Larry and Sergey's interests guided GoogleX. But when I joined, they decided they wanted more definition behind the lab's purpose. I asked Larry if he wanted me to build a research center."

"No," replied Page, "too boring."

"How about an innovation incubator?"

"Boring too."

So Teller thought for a while and finally asked, "So, are we taking moonshots?"

"That's it," answered Page, "that's exactly what we're doing."

And that is exactly what they're doing. Over the past few years, Google has repeatedly made headlines with the audacity of their moonshots, dedicating their skunk works to everything from space exploration and life extension to AI and robotics. In other words, as of right now, there is perhaps no other company in the world playing the skunk game at such an elevated level.

Over the next few pages, we're going to examine exactly how Google takes moonshots, giving you an inside look at their skunk methodology and paying attention to which of Kelly Johnson's initial ideas they've kept, which they've changed, and—from a psychological perspective—why.

Let's start with what they've kept the same.

On many levels, Google's moonshot factory is no different from a traditional skunk works. Isolation, for example, is also key to their process. "In any organization," says Teller, "the bulk of your people will be climbing the hill they're standing on. That's what you want them to do. That's their job. A skunk works does a totally different job. It's

a group of people looking for a better hill to climb. This is threatening to the rest of the organization. It just makes good sense to separate
these two groups."

It also makes sense to encourage skunk workers to take risks. "If
you're telling people to find a new mountain to climb," says Teller, "it's
pretty stupid to tell them to play it safe. Moonshots are risky. If you're
interested in tackling these challenges, you're going to have to embrace
some serious risk."

About this last bit—well, he's not kidding. Where GoogleX devi-
ates from traditional skunk is with the size of the goals they're setting.
Moonshots, by their definition, live in that gray area between audacious projects and pure science fiction. Instead of mere 10 percent
gains, they aim for 10x (meaning ten times) improvements—that's a
1000 percent increase in performance.

While a 10x improvement is gargantuan, Teller has very specific
reasons for aiming exactly that high. "You assume that going 10x bigger is going to be ten times harder," he continues, "but often it's literally easier to go bigger. Why should that be? It doesn't feel intuitively
right. But if you choose to make something 10 percent better, you
are almost by definition signing up for the status quo—and trying to
make it a little bit better. That means you start from the status quo,
with all its existing assumptions, locked into the tools, technologies,
and processes that you're going to try to slightly improve. It means
you're putting yourself and your people into a smartness contest with
everyone else in the world. Statistically, no matter the resources available, you're not going to win. But if you sign up for moonshot thinking, if you sign up to make something 10x better, there is no chance
of doing that with existing assumptions. You're going to have to throw
out the rule book. You're going to have to perspective-shift and sup-
plant all that smartness and resources with bravery and creativity."

This perspective shift is key. It encourages risk taking and enhances
creativity while simultaneously guarding against the inevitable decline.
Teller explains: "Even if you think you're going to go ten times bigger,
reality will eat into your 10x. It always does. There will be things that

will be more expensive, some that are slower; others that you didn't think were competitive will become competitive. If you shoot for 10x, you might only be at 2x by the time you're done. But 2x is still amazing. On the other hand, if you only shoot for 2x [i.e., 200 percent], you're only going to get 5 percent and it's going to cost you the perspective shift that comes from aiming bigger."

Most critically here, this 10x strategy doesn't hold true just for large corporations. "A start-up is simply a skunk works without the big company around it," says Teller. "The upside is there's no Borg to get sucked back into; the downside is you have no money. But that's not a reason not to go after moonshots. I think the opposite is true. If you publicly state your big goal, if you vocally commit yourself to making more progress than is actually possible using normal methods, there's no way back. In one fell swoop you've severed all ties between yourself and all the expert assumptions." Thus entrepreneurs, by striving for truly huge goals, are tapping into the same creativity accelerant that Google uses to achieve such goals.

That said, by itself, a willingness to take bigger risks is no guarantee of success. The rapid iteration approach of "fail early, fail often" still applies. As a result, some serious risk mitigation is equally critical.

At GoogleX, this mitigation comes from an especially vicious feedback process. "We try a lot of things," says Teller, "but we don't allow most of them to continue. At several different stages, we end most projects. Only a very small number are allowed to escalate to the next level. In the end, it looks like we got everything right, like we're geniuses— but that's not what's going on at all."

What's going on is data or death. GoogleX demands that all their projects be measurable and testable. They won't start a project if they don't have ways to judge its progress. And they do judge its progress— repeatedly. Sometimes projects end, sometimes they're absorbed into Google proper, sometimes they're stalled—meaning they can continue but are not allowed to grow. "As far as individual projects are concerned," says Teller, "this allows for a fairly freewheeling approach. But taken together, in aggregate, it's actually a fairly rigorous process."

It's also Darwinian evolution applied to rapid iteration. Big ideas for progress are competing against other big ideas for progress. While it's not a zero-sum game—as there's more than one winner—it's ruthless nonetheless. To put this in different terms, just as Google's version of skunk amps up the risk taking with the size of the goals they set (their 10x requirement), forcing those goals to compete in an experimental ecosystem amps up the risk mitigation.

And while average entrepreneurs might not be able to afford to start, stop, or stall dozens of projects at once—the Google ecosystem— they can set up multiple experimental tracks, employing rapid iteration and tighter feedback loops to fail forward far more consistently. Even better, as Teller argues, this kind of rigor brings a funding advantage. "People think that bold projects don't get funding because of their audacity. That's not the case. They don't get funded because of a lack of measurability. Nobody wants to make a large up-front investment and wait ten years for any sign of life. But more often than not, if you can show progress along the way, smart investors will come on some pretty crazy rides."

Google's Eight Innovation Principles

While Kelly Johnson had fourteen rules, Google has eight innovation principles that govern their strategy, famously summarized in a 2011 article by Google senior vice president of advertising Susan Wojcicki.[17] Throughout this book, we'll see them highlighted in different ways and exhibited by different people. Without doubt, these rules are core to your success as an exponential entrepreneur. My suggestion is that you write them on your wall, use them as a filter for your next start-up idea, but above all, don't ignore them. Let's take a quick look:

1. *Focus on the User.* We'll see this again in chapter 6, when Larry Page and Richard Branson speak about the importance of building customer-centric businesses.

2. *Share Everything.* In a hyperconnected world with massive amounts of cognitive surplus, it's critical to be open, allow the crowd to help you innovate, and build on each other's ideas.

3. *Look for Ideas Everywhere.* The entire third part of this book is dedicated to the principle that crowdsourcing can provide you with incredible ideas, insights, products, and services.

4. *Think Big but Start Small.* This is the basis for Singularity University's 10^0 thinking. You can start a company on day one that affects a small group, but aim to positively impact a billion people within a decade.

5. *Never Fail to Fail.* The importance of rapid iteration: Fail frequently, fail fast, and fail forward.

6. *Spark with Imagination, Fuel with Data.* Agility—that is, nimbleness—is a key discriminator against the large and linear. And agility requires lots of access to new and often wild ideas and lots of good data to separate the worthwhile from the wooly. For certain, the most successful start-ups today are data driven. They measure everything and use machine learning and algorithms to help them analyze that data to make decisions.

7. *Be a Platform.* Look at the most successful companies getting billion-dollar valuations . . . AirBnb, Uber, Instagram . . . they are all platform plays. Is yours?

8. *Have a Mission That Matters.* Perhaps most important, is the company you're starting built upon a massively transformative purpose? When the going gets hard, will you push on or give up? Passion is fundamental to forward progress.

Flow

In trying to parse the secrets to skunk, we've covered a motley crew of mind hacks. On the motivational side, we've explored bold goals, value-aligned bold goals, and the trifecta super-charge of autonomy, mastery, and purpose. On the performance side, we enhanced creativity with the perspective-shift of 10x, boosted risk taking with rapid iteration, and shortening learning cycles with fast feedback. Then, to derisk the whole process, we introduced the rigor of experimental eco-systems. But there's a larger point here—this crew is not so motley.

All of these mind hacks serve an additional function. Not only do they increase motivation and performance, they do double duty as triggers for the state of consciousness known as *flow*.[18]

Technically, flow is defined as an optimal state of consciousness where we feel our best and perform our best. And you've probably had some experience with this state. If you've ever lost an afternoon to a great conversation or become so involved in a work project that all else was forgotten, then you've tasted the experience. Flow describes these moments of total absorption, when we become so focused on the task at hand that everything else falls away. Action and awareness merge. Time flies. Self vanishes. All aspects of performance—mental and physical—go through the roof.

We call this experience *flow* because that is the sensation conferred. In the state, every action, each decision, leads effortlessly, fluidly, seam-lessly to the next. It's high-speed problem solving; it's being swept away by the river of ultimate performance.

This last bit is no exaggeration. Over a hundred and fifty years of research show that flow sits at the heart of almost every athletic cham-pionship, underpins major scientific breakthroughs, and accounts for significant progress in the arts. "In recent business studies," says John Hagel III,[19] cochairman of the Deloitte Center for the Edge, "top exec-utives report being five times more productive in flow." This is a stag-gering statistic. Five times more productive is a 500 percent increase. As Virgin CEO Richard Branson says, "In two hours [in flow], I can accomplish tremendous things . . . It's like there's no challenge I can't meet."[20]

Hagel explains further: "In all our studies of extreme performance improvement, the people and organizations who covered the most dis-tance in the shortest time were always the ones who were tapping into passion and finding flow."

How to find flow is a tricky question, yet it's one my coauthor, Steven Kotler, has spent the past fifteen years trying to answer. Steven is the cofounder and director of research for the Flow Genome Project, an organization dedicated to decoding the science of ultimate human performance.

One of the lessons to emerge from this work is that flow states have triggers—that is, preconditions that lead to more flow. There are seventeen flow triggers in total—three environmental, three psychological, ten social, and one creative. We'll go into greater detail about these triggers in the next section, but the first thing to know is that flow follows focus. It is a state of total absorption. Thus all seventeen triggers are ways of heightening and tightening focus, of driving attention into the now and thus driving flow.

This also brings us back to the secrets of skunk. All the various mind hacks described in this chapter—the so-called secrets—are also incredible focusing mechanisms. Increased risk taking is obvious: Flow follows focus, and consequences always catch our attention. As big goals have big consequences, they too serve this function. And value-aligned big goals work even better. When alignment exists, passion results. As we always pay more attention to those things we're passionate about, value-aligned big goals further increase focus. Autonomy, mastery, and purpose—which all serve to boost intrinsic motivation and further passion—do more of the same. Fast feedback, meanwhile, allows real-time course correction; thus we don't lose focus wondering about how to better our performance. By creating an environment packed with flow triggers, skunk works create a high-flow environment.

As a way of exploring how today's entrepreneur can create such an environment, I want to introduce the book's first how-to section. The idea here is to offer a series of actionable steps, immediately applicable to your life and work and guaranteed to move the needle. In this case, we're going to break down flow's seventeen triggers[21] in far more detail, focusing specifically on how they apply to exponential entrepreneurs.

Flow's Environmental Triggers

Environmental triggers are qualities in the environment that drive people deeper into flow.

High consequences are the first in this category. As mentioned above, flow follows focus, and consequences catch our attention. When there's danger lurking in the environment, we don't need to concentrate extra hard to drive focus; the elevated risk levels do the job for us.

And this doesn't just mean taking physical risks. The science shows that other risks—emotional, intellectual, creative, social—work just as well. "To reach flow," explains psychiatrist Ned Hallowell,[22] "one must be willing to take risks. The lover must be willing to risk rejection to enter this state. The athlete must be willing to risk physical harm, even loss of life, to enter this state. The artist must be willing to be scorned and despised by critics and the public and still push on. And the average person—you and me—must be willing to fail, look foolish, and fall flat on our faces should we wish to enter this state."

These facts also tell us that those exponential entrepreneurs with "fail forward" as their de facto motto have an incredible advantage. If people don't have the space to fail, then they don't have the ability to take risks. At Facebook, there is a sign hanging in the main stairwell that reads: "Move fast, break things." This kind of attitude is critical. If you're not incentivizing risk, you're denying access to flow—which is the only way to keep pace in a breakneck world.

Rich environment, the next environmental trigger, is a combination platter of novelty, unpredictability, and complexity—three elements that catch and hold our attention much like risk. Novelty means both danger and opportunity, and when either are present, it pays to pay attention. Unpredictability means we don't know what happens next; thus we pay extra attention to the next event. Complexity, when there's lots of salient information coming at us at once, does more of the same.

How to employ this trigger on the job? Simply increase the amount of novelty, complexity, and unpredictability in the environment. This is exactly what Astro Teller did by throwing out existing assumptions and demanding a 10x improvement. But it's also what Steve Jobs did when he designed Pixar. By building a large atrium at the building's center, then locating the mailboxes, cafeteria, meeting rooms, and most famously, the bathrooms, beside the atrium, he forced employees from all walks of the company to randomly bump into one another, massively increasing the amount of novelty, complexity, and unpredictability in their daily life.

Deep embodiment is a kind of total physical awareness. It means paying attention with multiple sensory streams at once. Take Montessori education. The Montessori classroom has been shown to be one of the highest flow environments on Earth.[23] Why? Because they emphasize learning through doing. Don't just read about that lighthouse, go out and build one. By working with your hands alongside your brain, you're engaging multiple sensory systems at once, grabbing hold of the attention system and forcing focus into the now.

Flow's Psychological Triggers

Psychological triggers are conditions in our inner environment that create more flow. They're psychological strategies for driving attention into the now.

Back in the 1970s, pioneering flow researcher Mihaly Csikszentmihalyi identified clear goals, immediate feedback, and the challenge/skills ratio as the three most critical.[24] Let's take a closer look.

Clear goals, our first psychological trigger, tell us where and when to put our attention. They are different than the high, hard problems of big goals. Those big goals refer to overarching passions: feeding the hungry, opening the space frontier. Clear goals, meanwhile, concern all the baby steps it's going to take to achieve those big goals. With these smaller goals, call them sub-goals, clarity is of the utmost importance for staying present and finding flow. When goals are clear, the mind doesn't have to wonder about what to do or what to do next—it already knows. Thus concentration tightens, motivation is heightened, and extraneous information gets filtered out. As a result, action and awareness start to merge, and we're pulled even deeper into now. Just as important, in the now, there's no past or future and a lot less room for self—which are the intruders most likely to yank us to the then.

This also tells us something about emphasis. When considering clear goals, most have a tendency to skip over the adjective *clear* to get to the noun *goals*. When told to set clear goals, we immediately visualize ourselves on the Olympic podium, the Academy Award stage, or the Fortune 500 list, saying, "I've been picturing this moment since I was fifteen," and think that's the point.

But those podium moments can pull us out of the present. Even if success is seconds away, it's still a future event subject to hopes, fears, and all sorts of now-crushing distraction. Think of the long list of infamous sporting chokes: the dropped pass in the final seconds of the Super Bowl; the missed putt at the end of the Augusta Masters. In those moments, the gravity of the goal pulled the participants out of the now, when, ironically, the now was all they needed to win.

If creating more flow is the aim, then the emphasis falls on *clear*, not *goals*. Clarity gives us certainty. We know what to do and where to focus our attention while we are doing it. When goals are clear, metacognition is replaced by in-the-moment cognition, and the self stays out of the picture.

Applying this idea in our daily life means breaking tasks into bite-size chunks and setting goals accordingly. A writer, for example, is better off trying to pen three great paragraphs at a time, rather than attempting one great chapter. Think challenging yet manageable—just enough stimulation to shortcut attention into the now, not enough stress to pull you back out again.

Immediate feedback, our next psychological trigger, is another shortcut into the now. The term refers to a direct, in-the-moment coupling between cause and effect. As a focusing mechanism, immediate feedback is something of an extension of clear goals. Clear goals tell us what we're doing; immediate feedback tells us how to do it better. If we know how to improve performance in real time, the mind doesn't go off in search of clues for betterment; we can keep ourselves fully present and fully focused and thus much more likely to be in flow.

Implementing this in business is fairly straightforward: Tighten feedback loops. Practice agile design. Put mechanisms in place so attention doesn't have to wander. Ask for more input. How much input? Well, forget quarterly reviews. Think daily reviews. Studies have found that in professions with less direct feedback loops—stock analysis, psychiatry, medicine—even the best get worse over time. Surgeons, by contrast, are the only physicians that improve the longer they're out of medical school. Why? Mess up on the table and someone dies. That's immediate feedback.

The challenge/skills ratio, the last of our psychological flow triggers, is arguably the most important. The idea behind this trigger is that attention is most engaged (i.e., in the now) when there's a very specific relationship between the difficulty of a task and our ability to perform that task. If the challenge is too great, fear swamps the system. If the challenge is too easy, we stop paying attention. Flow appears near the emotional midpoint between boredom and anxiety, in what scientists call the flow channel—the spot where the task is hard enough to make us stretch; not hard enough to make us snap.

This sweet spot keeps attention locked in the present. When the challenge is firmly within the boundaries of known skills—meaning I've done it before and am fairly certain I can do so again—the outcome is predetermined. We're interested, not riveted. But when we don't know what's going to happen next, we pay more attention to the next. Uncertainty is our rocket ride into the now.

Flow's Social Triggers

There is also a collective version of a flow state known as *group flow*.[25] This is what happens when a bunch of people enter the zone together. If you've ever seen a fourth-quarter comeback in football, where everyone is always in the right place at the right time and the result looks more like a well-choreographed dance than anything that normally happens on the gridiron—that's group flow in action.

But it's not just athletes who play this game. In fact, group flow is incredibly common in start-ups. When the whole team is driving toward a singular purpose with incredible speed—again, that's group flow in action. "Because entrepreneurship is about the nonstop navigation of uncertainty," says Salim Ismail,[26] "being in flow is a critical aspect of success. Flow states allow an entrepreneur to stay open and alert to possibilities, which could exist in any partnership, product insight, or customer interaction. The more flow created by a start-up team, the higher the chance of success. In fact, if your start-up team is not in a near-constant group flow state, you will not succeed. Peripheral vision gets lost and insights don't follow."

So how to precipitate group flow? This is where social triggers come into play. These triggers are ways to alter social conditions to produce more group flow. A number of them are already familiar. The first three—*serious concentration; shared, clear goals; good communication* (i.e., lots of immediate feedback)—are the collective versions of the psychological triggers identified by Csikszentmihalyi.

Two more—*equal participation* and an *element of risk* (mental, physical, whatever)—are self-explanatory given what we already know about flow. The remaining five require a little more information.

Familiarity, our next trigger, means the group has a common language, a shared knowledge base, and a communication style based on unspoken understandings. It means everybody is always on the same page, and when novel insights arise, momentum is not lost due to the need for lengthy explanation.

Then there's *blending egos*—which is kind of a collective version of humility. When egos have been blended, no one's hogging the spotlight and everyone's thoroughly involved.

A *sense of control* combines autonomy (being free to do what you want) and mastery (being good at what you do). It's about getting to choose your own challenges and having the necessary skills to surmount them.

Close listening occurs when we're fully engaged in the here and now. In conversation, this isn't about thinking about what witty thing to say next or what cutting sarcasm came last. Rather, it's generating real-time, unplanned responses to the dialogue as it unfolds.

Our final trigger, *Always say "yes, and . . . ,"* means interactions should be additive more than argumentative. The goal here is the momentum, togetherness, and innovation that comes from ceaselessly amplifying one another's ideas and actions. It's a trigger based on the first rule of improv comedy. If I open a sketch with "Hey, there's a blue elephant in the bathroom," and you respond with "No, there's not," the scene goes nowhere. Your denial kills the flow. But instead, if your response is of the "yes, and . . ." variety—"Yeah, sorry, I had no idea where to put him, did he leave the toilet seat up again?"—then the story goes someplace interesting.

Flow's Creative Trigger

If you look under the hood of creativity, what you see is pattern recognition (the brain's ability to link new ideas together) and risk taking (the courage to bring those new ideas into the world). Both of these experiences produce powerful neurochemical reactions and the brain rides these reactions deeper into flow.

This means, for those of us who want more flow in our lives, we have to think different, it's as simple as that. Instead of tackling problems from familiar angles, go at them backward and sideways and with style. Go out of your way to stretch your imagination. Massively up the amount of novelty in your life; the research shows that new environments and experiences are often the jumping-off point for new ideas (more opportunity for pattern recognition). Most important, make creativity a value and a virtue. This is where we return to moonshot thinking again. As Teller explains, "You don't spend your time being bothered that you can't teleport from here to Japan, because there's a part of you that thinks it's impossible. Moonshot thinking is choosing to be bothered by that."

Final Advice

One of the most well-established facts about flow is that the state is ubiquitous—meaning it shows up anywhere, in anyone, provided certain initial conditions are met. What are these conditions? These seventeen triggers—it really is that straightforward.

And there's a reason for this as well. We're biological organisms, and evolution is conservative by design. When a particular adaptation works, the basic design is repeated again and again. Flow most certainly works. As a result, our brains are hardwired for the experience. We are all designed for optimal performance—it's a built-in feature of being human.

CHAPTER FIVE

The Secrets of Going Big

Born Above the Line of Super-Credibility

"I watched the news today and I saw something sooooo . . . awesome," says Jon Stewart, host of Comedy Central's *Daily Show*.[1] It's April 24, 2012, and Stewart is, well, a little excited. His eyebrows dart, his nostrils flare, he's about to blow. A newsreel begins to roll. We see an anchorman in a suit, hands folded, cucumber calm: "This may seem like science fiction," he says, "but today a group of space pioneers announced plans to mine asteroids for precious minerals." Cut back to Stewart, in a tizzy, shouting, "Space pioneers going to mine asteroids for precious materials! BOOM! BOOM! YES! Stu-Beef is all in. Do you know how rarely the news in 2012 looks and sounds like you thought news would look and sound in 2012?"

What Stewart was boom-booming about was Planetary Resources, Inc.,[2] the asteroid-mining company I cofounded with Eric Anderson in 2009 and announced in 2012. Clearly, asteroid mining is a crazy science-fiction idea, bold on every level. To start this kind of company with any real hope of success and—equally difficult—to present it to the public in a plausible fashion requires a different kind of approach. Over the years, I've developed a series of strategies for tackling these kinds of challenges, none more important than birthing projects above the line of super-credibility.

As will become clearer later, getting above that line requires a deep passion. Mine emerged in 1969. I was only eight years old when Apollo 11 landed on the moon, and I decided then and there that going into space was what I wanted to do with my life. I was in my early twenties when I realized NASA was never going to get me there. Constrained by government spending and frightened by the risk of failure, the space agency had become a military-industrial jobs program unlikely to return to the Moon or push onward to Mars. It was clear to me, if we were going to boldly go, it was going to have to be without the help of government.

Thus I devoted the next thirty years to starting private ventures that I thought would open the space frontier. In addition to my work with the International Space University, these included three efforts to jump-start a space-tourism economy: the XPRIZE, Zero-G, and Space Adventures Limited.

It was thru the founding of Space Adventures[3] that I teamed up with Eric Anderson, my future partner in Planetary Resources. Back in 1995, Anderson, a recent University of Virginia aerospace graduate, joined me as an intern to help develop a company to leverage the vast assets of the once-powerful Soviet space program, now hungry for hard currency and willing to offer anyone with enough cash a ride into space. Within a year, Anderson worked his way up from intern to vice-president to president of Space Adventures (later CEO). Over the next fifteen years, he took the company to over $600 million in cumulative revenues—and if you've ever tried to sell a $50 million seat into orbit or a $150 million ride around the Moon, you'll respect the achievement.

Others, impatient for this same dream, took a different approach. On October 4, 2004, when aviation legend Burt Rutan won the $10 million Ansari XPRIZE with SpaceShipOne, Sir Richard Branson swooped in to license the winning technology, committing a quarter of a billion dollars to develop Virgin Galactic's SpaceShipTwo—the commercial follow-up to SpaceShipOne.[4] Next, Amazon founder Jeff Bezos committed over $100 million toward a secretive launch vehicle company called Blue Origin.[5] Perhaps most impressive was PayPal cofounder-turned-aerospace-disrupter Elon Musk, whose epic success with the Falcon 9 launch vehicle and Dragon capsule placed him in the category of "space god" and earned him a multibillion-dollar contract from NASA to ferry cargo to the International Space Station.[6]

Certainly, these were all amazing successes, but, in the summer of 2009, when I got together with Eric for our annual "What's Next?" retreat, our outlook on the future of space was gloomy at best. Despite these great wins, everything was still moving too slowly. To really open up the space frontier we needed more than a dozen people heading into orbit—we needed hundreds of thousands. We needed the advantages of scale.

It had become clear to us that if we were ever going to open up space, then we needed to exploit the same economic engine that had opened every previous frontier: the search for resources. "Whether it was the Chinese pioneering the Silk Road or early European explorers looking for gold and spices on the other side of oceans or American settlers scouring the West for timber and land," says Anderson,[7] "the search for new resources has always been the main reason for attempting the difficult and dangerous." This was when Eric and I started having a serious discussion about asteroid mining.

We weren't the first folks to have this discussion. The idea of catching up to giant floating rocks, mining them for precious metals and ores, and hauling that loot back to Earth dates back to 1895, when first proposed by the father of the Russian space program, Konstantin E. Tsiolkovsky.[8] Between the nineteenth century and the twenty-first, asteroid mining became a science-fiction mainstay, but it started to become science fact in the 1990s, when a trio of space missions (NASA's NEAR Shoemaker and Stardust, and the Japan Aerospace Exploration Agency's Hayabusa) all managed to catch up to asteroids (and two managed to scrape the surface of these and bring back microscopic samples).[9] But the gap between those science missions and the full-scale industrial efforts that Eric and I were dreaming about was massive—and that is exactly the point.

To pull off such a massive moonshot, we're going to need help, a *lot* of help. And thus our first challenge—convincing anyone our dream was doable. This meant, for certain, we were going to have to give birth to this dream above the line of super-credibility.

Let me explain: In each of our minds we have a line of credibility. When you first hear a new idea, you place it above or below this line. If you place it below, you dismiss it immediately, often as ridiculous. If you place it above, you're willing to give it the benefit of the doubt, follow it over time, and continue to make serial judgments. But we also have a line of *super*-credibility. When a new idea is born above this line, you accept it immediately and say, "Wow, that's fantastic! How can I get involved?" The idea is so convincing that your mind accepts it as fact and your focus shifts from probabilities to implications.

Plotting the Line of Super Credibility

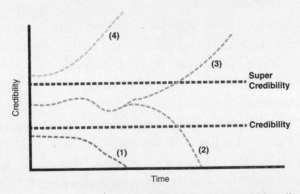

Plotting the Line of Super-Credibility. (1) Non-credible rollout; (2) credible rollout, non-credible performance; (3) credible rollout, super-credible performance; and (4) super-credible rollout.

Source: Peter H. Diamandis

Unless Planetary Resources was introduced to the world far above that line, clearly it would be dismissed out of hand. We needed to assemble a team that people would intuitively trust to execute this vision. Chris Lewicki—who had run three different billion-dollar Mars missions at NASA's fabled Jet Propulsion Laboratory (JPL)—was our first stop. With him as our president and chief engineer, we went on to recruit many of the top engineers who built, designed, and operated Mars rover *Curiosity* (we knew we were on the right track when Eric received a call from the head of JPL asking us to kindly stop recruiting his best people).

And had we stopped there, we might have launched in a credible fashion. We were certainly believable. Both Eric and myself are respected members of the space community. Our team was an assortment of the best and the brightest. But because we were proposing to do something as bold as asteroid mining, credible wasn't enough.

For this reason, we kept the company secret for nearly three years, spending that period pushing ourselves further toward the line of super-credibility. To that end, we recruited a bevy of billionaire investors willing to put their cash and their names behind the project. These were folks like Larry Page, Eric Schmidt, Ram Shriram (Google's first investor), Ross Perot, Jr., Charles Simonyi (Microsoft's chief architect), and Richard Branson. Recruiting such big names gave us a number of advantages. For starters, just getting through the gauntlet of their due diligence meant getting the benefit of their thinking. Having the smartest folks on the planet pound on your vision can help turn dank coal into glittering diamonds. More important, when we finally did launch, these names drew a crowd. And that's the bigger point. It's hard to argue with the combination of the planet's top space engineers and most respected businessmen, so we entered the public eye far above the line of super-credibility. Which is exactly why the news in 2012, as Jon Stewart pointed out, sounded exactly like you thought the news would sound in 2012.

The International Space University

Of course, right now, you're probably thinking this super-credibility advice isn't much good for entrepreneurs without billionaires in their Rolodex. Certainly, when Eric and I started on our investor recruitment mission, we already had a network in place that gave us access to investors like Branson and Page. This is not going to be the case for everyone. But that doesn't mean all is lost. In fact, my entire thinking about the line of super-credibility dates back to a time in my life when I had little credibility, when I was a college student—in the pre-Internet, pre-Google, pre-Facebook days—with access to few beyond friends and family.

This story starts in 1980, during my sophomore year at MIT, when I founded Students for the Exploration and Development of Space (SEDS).[10] SEDS emerged from my passion to open the space frontier and my frustration—already mentioned—with NASA. Alongside early SEDS leaders and fellow "space cadets" Bob Richards and Todd Hawley,[11] we stitched together an organization of thirty college chapters from around the world that were all committed to promoting student participation in space. In 1982, because of SEDS, the three of us were invited to Vienna, Austria, to present at the United Nations Committee on the Peaceful Uses of Outer Space. It was there we met and befriended Arthur C. Clarke—author of *2001: A Space Odyssey* and inventor of the geostationary satellite.

"Uncle Arthur," as we called him, shared stories about how close the space inventors, engineers, and visionaries of the 1940s and 1950s were to one another. It was the power of their collective friendship, knowledge, and vision that ultimately gave birth to the Apollo Program. The idea of such a tight network got us dreaming about creating an International Space University, or ISU, a place where the space cadets of today could dream up tomorrow. Even bolder, we imagined having our campus in orbit, an off-world university where students could live, study, and do research.[12]

Of course, for such an institution to come into existence, people were going to have to believe it was viable and worthwhile. And because we were just graduate students, being credible wasn't enough. We needed to be super-credible. So how did we do it? Here's our playbook, laid out one step at a time.

Step One: Familiarity matters. We started by recruiting the help of people who had seen us succeed over the previous five years with SEDS. This may sound obvious, but in meeting hundreds of entrepreneurs over the years, I've discovered that many of them have forgotten the obvious: The very best people to help you with your next project are those who helped you or watched you succeed with your last.

In the start-up game, especially early in your career, backers are typically close friends and family, the people who already know and trust you. Once you've moved beyond that circle—or if you don't happen to have that circle—the folks most likely to invest in your success are those who have already watched you succeed. So if you're lacking a track record, make one. Start your bold project with a much smaller effort aimed at letting others see you pull it off. Then tap that network for your next step. For sure, we could never have pulled off ISU without first having succeeded with SEDS.

Step Two: Slow down and build credibility. Instead of rushing headlong toward our bold goal of a space university, our first step was to organize a conference to "study" the feasibility of a space university. Many entrepreneurs skip this step. They have a bold idea, get a little traction, and mistake that vote of confidence for a sign that big dollars are around the corner. Perhaps, but real traction means more than just a little confidence; it requires a lot of trust. Investors love ideas, but they fund execution. And for us, well, a conference was already something we knew how to run.

Over the course of a few months we managed to raise $50,000—mostly around the idea of holding an aerospace "jobs fair" at MIT in parallel with our ISU feasibility conference.[13] Our big break came when Bob Richards—who was then living in Toronto—managed to get the commitment of the head of the Canadian Space Agency (CSA) to come down and speak. Then we leveraged our luck. With CSA attending, we were able to convince the European Space Agency (ESA) to attend, then based on having both CSA and ESA, the Japanese agreed to join, followed by the Russian Federal Space Agency, the Chinese National Space Administration, the Indian Space Research Organization, and finally even NASA. Slowly, bit by bit, we were climbing toward super-credibility.

But there was a ways to go. Thus, onto Step Four: Messaging matters.

In the six months before the conference, the three of us brainstormed what a space university would actually look like, drilling deep into the particulars of what we would teach and who would attend. We also developed a detailed plan and ceaselessly bounced it off of our advisers—getting as much engagement as possible. This engagement mattered so much because getting our plan adopted depended both on the quality of our ideas and—the bigger point—on who presented those ideas to the world. At our conference, rather than present the plan ourselves, we asked our advisers to do the talking for us. Dr. Byron Lichtenberg, two-time Space Shuttle astronaut and cofounder of the Association of Space Explorers, presented the academic plan. Dr. John McLucas, past secretary of the US Air Force and CEO of Comsat, presented the financing plan, and Dr. Joseph Pelton, then head of international affairs for INTELSAT, presented our governance plan.

And it worked. So credible were both the ideas and the folks presenting them that we jumped far above the line of super-credibility. An event that was billed as a conference to study the idea of ISU quickly became the ISU Founding Conference. So how far can super-credibility take those with little credibility? Before the weekend was out, we'd raised the seed capital to launch a summer program—our first step on the road to an actual university.

Staging Toward Bold

There's a little more to the ISU story, and we'll get there in a moment, but first I wanted to pause and consider a second lesson that we learned on our way to founding that university: the critical importance of staging your bold ideas. In the six months before that founding conference, when Bob, Todd, and I were drilling down into space school particulars, what we were really doing was breaking our vision into executable, bite-size chunks, what psychologists call subgoals.

Subgoals bring dual benefits. The first is the alignment of risk with reward. Few projects ever receive all the funding they need at the beginning. Usually, capital comes in stages as entrepreneurs find new ways to mitigate risk. Instead of one lump sum, money arrives in discrete waves: seed capital, crowdsourced capital, angel capital, super-angel capital, strategic partners, series A venture, series B venture, and sometimes even a public offering. More and more investment comes as each increment proves the capability of the management team and the veracity of the vision.

The second benefit to subgoals is psychological. In the last chapter, we met Gary Latham and Edwin Locke and learned that there's hidden leverage in setting big goals. But we also learned that this is true only when certain "if-then" conditions are satisfied. Commitment—meaning the alignment of values and goals—was merely the first of these. Equally important is confidence.

"Big goals only increase motivation," explains Latham,[14] "when the person setting those goals is confident in their ability to achieve them. This means breaking big goals apart into achievable subgoals."

It's for these reasons that in the six months leading up to the ISU Founding Conference, we broke down our moonshot into five executable steps:

1. Hold a conference at MIT to study the idea of ISU.
2. Borrow the campus of MIT to hold a nine-week ISU summer session and invite a hundred graduate students to participate.

3. Repeat the same summer program in additional countries to prove out the concept and build a global community.
4. Establish a permanent terrestrial campus.
5. Establish an orbital space campus on the International Space Station.

While steps four and five—our moonshot goals—were aimed at capturing our supporters' hearts, steps 1, 2, and 3, being much more incremental (thus believable) were aimed at their minds. And it worked. After receiving this first bit of support, we went from building our team to holding our feasibility conference to launching our first summer session.

The session was magical, gathering a hundred and four graduate students from twenty-one countries. Sure, it was totally bootstrapped—our campus borrowed, our faculty on loan (made up of the professors Bob, Todd and I recruited and borrowed from our respective alma maters). Yet it was still a complete success

Then we did it again, changing only the location (so we could create engagement in wider and wider communities). During that second summer, ISU borrowed the Université Louis Pasteur campus in Strasbourg, France. Then we were off to Toronto, Canada, in 1990, Toulouse, France, in 1991, and Kitakyushu, Japan, in 1992.

After the university had five years and about 550 alumni under its belt, we finally decided to try and parlay our assets into step 4 of our vision—a permanent terrestrial campus. One small problem: We had no tangible assets. As a fully virtual university with no campus, no cash, and a borrowed faculty—our only assets were our brand, our alumni, and our vision. Thus it was time to make stone soup.[15]

How to Make Stone Soup

A long time ago, in a tiny medieval village, a farmer spots three soldiers on the edge of town. Knowing what would likely happen next, he runs into the marketplace shouting a warning: "Quick, close the doors, lock the windows! There are soldiers coming and they'll take away all our food."

The soldiers are in fact hungry. When they enter the village, they start knocking on doors, asking for food. The first villager tells them the cupboard is bare. At the next, the second villager tells them the same. The next door isn't even opened.

Finally one of the starving soldiers says, "I have an idea—let's make stone soup."[16]

With that, he strides over and knocks on yet another door. "Excuse me," he says to the villager, "do you have a cauldron and some firewood? We would like to make some stone soup."

The villager, thinking there's no risk to her, says, "Soup from stones? This I've got to see. Sure, I'll help." So she gives them a cauldron and some firewood while another soldier gets some water. They bring the water to a boil and place three large stones in the pot. News spreads around the town and the villagers begin to gather. "Soup from stones," they say. "This we have to see."

So the soldiers are standing around the fire and the villagers are standing around the soldiers.

"I had no idea you can make soup from stones," says one villager.

"Sure can," replies the soldier.

Eventually, tired of standing around, another villager asks, "Can I help?"

"Perhaps," says a soldier, "if you had a few potatoes, that would make the stone soup even better."

So the villager quickly fetches some potatoes and adds them to the pot of simmering stones.

Another asks, "How can I help?"

"Well, a couple of carrots would sure make the soup even better."

So the villager contributes some carrots. Soon others are adding poultry, barley, garlic, and leeks. After a while one of the soldiers calls out, "It's done," and shares the soup with everyone. The villagers are heard saying, "Soup from stones! It tastes fantastic. I had no idea."

That story of stone soup comes from an old folktale that eventually became a children's book. I heard it in college and it's never left me. In fact, I've come to think of making stone soup as the only way an entrepreneur can succeed. The stones are, of course, your big bold ideas; the contributions of the villagers, the capital, resource, and intellectual support offered by investors and strategic partners. Everyone who adds a small amount to your stone soup is in fact helping to make your dreams come true.

What makes stone soup work is passion. People love passion. People love to contribute to passion. And you can't fake it. The human bullshit detector is great at spotting the inauthentic article. The used car salesman, the carnival barker, and the disingenuous politician always rub us wrong.

Sure, I'm probably not telling you something you don't already know. But passion is a trickier subject than most assume. For starters, there are versions of passion that are extremely unhelpful to entrepreneurs, such as what John Hagel III, the cofounder of Deloitte's Center for the Edge, calls the passion of the true believer. "In Silicon Valley we have many examples of the true believer," says Hagel.[17] "These are great entrepreneurs [who] are truly passionate about a very specific path and are notoriously not open to alternative views or approaches. Their passion is enduring and it does focus, but it can also be blind—leading the entrepreneur to reject critical input that does not match their preconceived views."

Hagel and colleagues have made quite a study of passion,[18] coming to define the version that best serves individuals and organizations as the passion of the explorer. "These are people who see a domain," he continues, "but not the path. The fact that the path is not clearly defined is what excites them and motivates them. . . . It also makes them alert to a variety of inputs that can help them to better understand the domain and discover more promising paths. . . . [Thus] they are constantly balancing the need to move forward with the need in the moment to reflect on their experiences."

This is the same kind of passion that makes stone soup. Passionate people are deeply creative in seeking out and pulling in the resources they need to pursue their passion, but it goes further than that. "People who pursue their passions inevitably create beacons that attract others who share their vision," says Hagel. "Few of these beacons are consciously created; they are by-products of pursuing one's passion. Passionate people share their creations widely, leaving tracks for others to find them."

And this is exactly what happened with ISU. In 1992, as we sought to establish our permanent terrestrial campus, we put out an RFP (request for proposals) that basically said, "Hi there, we're ISU. We have this concept for a permanent campus. We've held five summer programs in five different cities, and this is our vision for what we want to create and where we want to go. Please tell us how much cash endowment, buildings, and operational money you will give us to bring our vision to your city."

Had we gotten no response at all, I would not have been surprised. But that wasn't the case. Within six months, we received seven proposals ranging from $20 million to $50 million in funding, buildings, faculty, equipment, and even the promise of accreditation. In short, everything we needed to implement the next phase of ISU.

So how far can the right combination of super-credibility, staged goals, and passion take you? In our case, pretty far. In the end, the city of Strasbourg, France, won the bid. They went on to build us a beautiful $50 million campus in Parc d'innovation. Today many of the heads of the world's space agencies are ISU alumni. And while we haven't yet built our orbital extension, we're definitely betting that once asteroid mining becomes the norm, our space-based campus won't be far behind.

Peter's Laws—Mindset Matters

During the earliest days of ISU, I shared an office with Todd Hawley, who as a joke put a copy of Murphy's Law on the wall. That depressing advice—"If anything can go wrong, it will"—stared at me every day. It also started to get under my skin. There's an old saying in business: You're the average of the five people you spend the most time with. The same is true for ideas. As was pointed out in the last chapter, mindset matters. Thus, a week into Murphy's mental assault, I went to the whiteboard behind my desk and wrote: "If anything can go wrong, fix it! (To hell with Murphy!)" Then above the quote I wrote, "Peter's Law."

Over the years that followed I started collecting more laws— principles and truisms that have guided me in times of difficulty and opportunity. Most of these are my fundamental rules to live by, my go-to principles when the proverbial shit hits the fan. In the rest of this section, we'll take a closer look, but before we get to my ideas, we first need to address something far more important: your ideas.

The maxims presented below are the ones that have worked for me, but that's no guarantee they'll work for you. So come up with your own. Borrow from anyone you like. The point isn't to produce pretty pictures covered with inspirational quotes. The point is to trust your history. Plumb your past to plot your future. Start collecting mind hacks by examining your own life and seeing what strategies *consistently* worked along the way. Turn those strategies into your laws.

Why is this so important? Because fear is hell on decision making. As threat levels begin to rise, the brain starts limiting our options. The fight-or-flight response is the extreme version of this story. When we are confronting mortal terror, our choices are literally limited to three: fight, flight, or freeze. But the same thing starts to happen with lesser fears. As Emory University neuroeconomist Gregory Berns wrote in an article for the *New York Times*:[19] "Fear prompts retreat. It is the antipode of progress." And that's why it's important to write down your own laws. You're essentially creating an external hard drive for when your internal hard drive is guaranteed to crash.

At the end of this chapter, you'll find a full list of my laws, but first here are a few favorites, with some story and explanation to back them up and hopefully make them more useful. Steal from me, borrow from others, modify at will, but most important, take action and create a list of your own.

#17: THE BEST WAY TO PREDICT THE FUTURE IS TO CREATE IT YOURSELF!

I've seen variations on this quote attributed to everyone from Abraham Lincoln to Peter Drucker—which certainly makes it enduring. And for good reason. The future is not preordained. It unfolds as a result of action—the choices you make and the risks you take. At a very fundamental level, this is exactly what it means to be an entrepreneur. Have a vision for tomorrow, pull yourself toward it. I wanted a future that included private commercial space flight, so I launched the XPRIZE. I saw asteroid mining as a viable reality, so I cofounded Planetary Resources.

#10: WHEN FACED WITHOUT A CHALLENGE—MAKE ONE!

We humans are hardwired for challenge. This is why flow—the state of optimal human performance—shows up only outside of our comfort zone, when we are pushing limits and using skills to the utmost. Want proof? How about the significant correlation between early retirement and death. According to a 2005 report in the British Medical Journal, people who retire at fifty-five are 89 percent more likely to die in the ten years after retirement than those who retire at sixty-five. We need to be alive to stay alive, simple as that.

#11: NO SIMPLY MEANS BEGIN ONE LEVEL HIGHER!

When someone says no, it's often because they're not empowered to say yes. In many organizations, the only person who can say yes is the one atop the food chain. When I was in graduate school, I was desperate for a ride on NASA's zero-gravity (parabolic flight) airplane. I tried everything to get aboard (including volunteering as a medical guinea pig), but could never get permission. So I took things to the next level, partnering with two friends to start a commercial company (Zero-G) to offer this same service.

But getting permission to start this service took a while—eleven years, to be exact. Over the next decade, we battled an army of FAA lawyers who all insisted that large-scale commercial zero-g operations were not possible under the federal aviation regulations, despite the fact that NASA had been operating parabolic flights for over thirty years. They kept demanding that I show them where in the regulations it says an airplane is allowed to fly parabolic arcs. I had only one answer: "Show me where it says I can't." Quite simply, none of these midlevel bureaucrats had the power to say yes. Finally, a decade later, my request made it all the way up to the FAA administrator, Marion Blakey, an amazing woman who had the right answer: "Of course you should be able to do this—let's figure out how."

#1: IF ANYTHING CAN GO WRONG, FIX IT! (TO HELL WITH MURPHY!)

Back in 2007, I decided that the world's foremost expert on gravity deserved the opportunity to experience zero gravity, so I offered professor Stephen Hawking a parabolic flight. He accepted, and we issued a press release. This is when our friends at the FAA—whose unofficial motto is clearly "we're not happy until you're not happy"—reminded us that our operating license permitted us to fly only "able-bodied" passengers, and Hawking, being totally paralyzed and wheelchair bound, did not qualify.

But to hell with Murphy. I decided to fix the problem. First, we had to determine who—in the FAA's mind—decides that someone is able-bodied? Second, if we could get that someone to declare Hawking "able-bodied," we still had to derisk our moonshot and ensure his safety.

After wading through lawyers, we determined that only Hawking's personal physicians and perhaps experts from the space-medicine world were qualified to make that call. So after purchasing malpractice insurance policies for a few of these folks, we were able to submit three letters to the FAA stating, without question, that Hawking was fit for a flight.

To address the second challenge, we decided that four physicians and two nurses would accompany him on the trip, then assembled a flying emergency room on board the airplane and flew a lengthy practice flight, training the medical team for everything from a heart attack to broken bones. We also decided (and announced to the public) that we'd fly a single thirty-second parabola, and maybe, if everything went perfectly, a second one.

At least that was the plan. The problem with the plan was Hawking. Not only did he endure that first parabola, he had—as he told me—the best time of his life. So we flew another and another and still he wanted more. In total, we made eight arcs with him aboard. Then, on the heels of this success, we had the amazing opportunity to fly six wheelchair-bound teenagers into zero gravity. These were kids who had never walked a day in their lives, yet they got to soar like Superman on that flight. The moral of the story: Stuff goes wrong. Expect it, learn from it, fix it—that's how remarkable happens.

#2: WHEN GIVEN A CHOICE—TAKE BOTH!

We're taught that when you are given a choice you have to choose only one option. But why choose? All through graduate school I was told to either go to school or start a company. It was binary or bust. But not for me. In my case, the answer was both and then some. I started three companies while in graduate school. I started eight more before I was forty. Steve Jobs juggled both Apple and Pixar. Elon Musk runs three multibillion-dollar successes: Tesla Motors, SpaceX, and SolarCity. Branson, well, alongside his Virgin Management group, has started over five hundred companies, including eight billion-dollar companies in eight different industries. This multiple-choice approach—if properly managed—can create tremendous momentum. Ideas cross-pollinate. Networks expand. The whole becomes much bigger than the sum of its parts.

#18: THE RATIO OF SOMETHING TO NOTHING IS INFINITE

The best predictor of future success is past action. It doesn't matter how small those actions. When I'm interviewing potential employees, I'm always more interested in what they've done than in what they will do. Doing something, doing anything, is always so much more important than just talking about doing it. The ratio of something to nothing is literally infinite. So make a plan. Set subgoals. Get busy. Even if the path is unclear, you'll use what you've learned taking that first step to build toward the next, and the next after that. Results always follow. Charles Lindbergh was correct: "The important thing is to start; to lay a plan, and then follow it step by step no matter how small or large one by itself may seem."

#21: AN EXPERT IS SOMEONE WHO CAN TELL YOU EXACTLY HOW SOMETHING CAN'T BE DONE

When I first dreamed up the XPRIZE, I went to all the major aerospace contractors looking for funding. They were dismissive. When the prize was announced, these same experts derided it. It took only eight years for Burt Rutan to prove them wrong. Henry Ford, when asked about his employees, said it best: "None of our men are 'experts.' We have most unfortunately found it necessary to get rid of a man as soon as he thinks himself an expert because no one ever considers himself expert if he really knows his job. A man who knows a job sees so much more to be done than he has done, that he is always pressing forward and never gives an instant of thought to how good and how efficient he is. Thinking always ahead, thinking always of trying to do more, brings a state of mind in which nothing is impossible. The moment one gets into the 'expert' state of mind a great number of things become impossible."

Peter's Laws™

The Creed of the Persistent and Passionate Mind

1. If anything can go wrong, fix it! (*To hell with Murphy!*)
2. When given a choice—*take both!*
3. Multiple projects lead to multiple successes.
4. Start at the top, then work your way up.
5. Do it by the book . . . *but be the author!*
6. When forced to compromise, ask for more.
7. If you can't win, change the rules.
8. If you can't change the rules, then ignore them.
9. Perfection is not optional.
10. When faced without a challenge—make one.
11. No simply means begin one level higher.
12. Don't walk when you can run.
13. When in doubt: *THINK!*
14. Patience is a virtue, but persistence to the point of success is a blessing.
15. The squeaky wheel gets replaced.
16. The faster you move, the slower time passes, the longer you live.
17. The best way to predict the future is to create it yourself!
18. The ratio of something to nothing is infinite.
19. You get what you incentivize.
20. If you think it is impossible, then it is *for you.*
21. An expert is someone who can tell you exactly how something can't be done.

22. The day before something is a breakthrough, it's a crazy idea.
23. If it was easy, it would have been done already.
24. Without a target you'll miss it every time.
25. Fail early, fail often, fail forward!
26. If you can't measure it, you can't improve it.
27. The world's most precious resource is the persistent and passionate human mind.
28. Bureaucracy is an obstacle to be conquered with persistence, confidence, and a bulldozer when necessary.

* Laws 12 and 15 by Todd B. Hawley. Law, 17 adopted from Alan Kay, Law 21 adopted from Robert Heinlein, Law 24 by Byron K. Lichtenberg, Law 25 adopted from John Maxwell.

CHAPTER SIX

Billionaire Wisdom

Thinking at Scale

Four Who Changed the World

In the past five chapters, we've been examining ways to raise your game. Exponential technologies added physical leverage, psychological tools provided a mental edge, and the combination allows entrepreneurs to become true forces for disruption. This chapter, which marks the end of that psychological exploration, focuses on the mind hacks of four remarkable men, a quartet of entrepreneurs who have already harnessed exponential technology to build multibillion-dollar companies that forever changed the world: Elon Musk, Richard Branson, Jeff Bezos, and Larry Page.

I've had the chance to work in varying degrees with each of these men. Elon Musk and Larry Page are both trustees and benefactors of the XPRIZE Foundation; Jeff Bezos ran the SEDS chapter at Princeton and has been passionate about opening space for the past forty years; and Richard Branson licensed the winning technology resulting from the Ansari XPRIZE to create Virgin Galactic. All four exemplify the central idea in this book, exhibiting a commitment to bold that's fierce, enduring, and masterfully executed.

Equally important, each of these entrepreneurs mastered a rarely discussed skill fundamental to bold pursuits and exponential entrepreneurship: the ability to think at scale. Exponential technology allows us to scale up like never before. Small groups can have huge impacts. A team of passionate innovators can alter the lives of a billion people in an eyeblink. To say that this kind of impact is unfathomable is putting it mildly.

Humans don't grok scale. Our brains evolved to process a simpler world, where everything we encountered was local and linear. Yet the four men described in this chapter have pushed through the limitations of linear thinking, and understanding their strategies for thinking at scale can help us do the same.

To get at those strategies, besides multiple conversations over years and in-person interviews with each, my research team combed through over two hundred hours of video footage of these men, breaking their ideas into categories and analyzing from there. In the end, we discovered that to think at scale, all four have leaned heavily on three of the psychological tools covered in earlier chapters—though, as we'll see in a moment, each in different and insightful ways—and, equally crucial, each relies on five additional mental strategies. While we'll get into greater detail in a moment, here's a complete list:

1. Risk taking and risk mitigation
2. Rapid iteration and ceaseless experimentation
3. Passion and purpose
4. Long-term thinking
5. Customer-centric thinking
6. Probabilistic thinking
7. Rationally optimistic thinking
8. Reliance on first principles, aka fundamental truths

Now, to see how these strategies work their magic, let's meet the first of our entrepreneurs, a man who reinvented banking before he turned thirty and then really got down to business.

Elon Musk and Life on Mars

In the early days of *Iron Man*, director Jon Favreau had a plausibility problem. His main protagonist, billionaire-genius-superhero Tony Stark, was larger than life. Too much larger than life. "I had no idea how to make him seem real," Favreau told *Time*.[1] "Then Robert Downey Jr. said, 'We need to sit down with Elon Musk.' He was right."

While Musk has yet to fabricate an Iron Man suit, he has revolutionized industries and built four multibillion-dollar companies: PayPal (banking), SpaceX (aerospace/defense), Tesla Motors (automotive), and SolarCities (power generation). As you might expect, with this kind of résumé, his passion for entrepreneurship emerged early.[2]

Born in Pretoria, South Africa, Musk was programming computers by age nine. At twelve, he made five hundred dollars selling the code for a video game called Blastar. He entered college at seventeen, spending two years at Queen's University in Ontario, Canada, before transferring to Wharton to study business and physics. Next Musk moved on to Stanford to pursue his PhD in applied physics. That pursuit didn't last long. He left the program after just two days—itching to jump into the exploding world of the Internet.

"My initial goal wasn't to start a company," he explains.[3] "I actually tried to get a job at Netscape—which seemed to be the only interesting Internet company at the time. I sent them my résumé, even hung out in the lobby, but I was too shy to talk to anyone and they never offered me a job. Eventually I said the hell with it, began coding myself, and started Zip2."

Zip2 was an application that allowed companies to post content—maps, directory listings, etc.—online. Back then, this was a neat enough trick that Musk's start-up was bought by Compaq for $307 million—which was then the biggest sum ever paid for an Internet company. Musk cleared his first $28 million.[4]

Next came X.com, an online financial services company that would eventually change its name to PayPal. Only three years later PayPal was sold to eBay for $1.5 billion in stock. Musk, the largest shareholder, walked away with his first $100 million.[5] Now the question was, what next?

"In early 2001, my old college roommate, Adeo Ressi, asked what I was going to do after PayPal. I remember telling him that there were certain things I thought were important for humanity's future—the Internet, sustainable energy, and space. At that time, after PayPal, I was very interested in space—which I believed was something clearly in the domain of governments—but that conversation with Adeo got me wondering when NASA was planning to send humans to Mars. I went looking, but couldn't find an answer on their website. At first I thought NASA just had a badly designed website. Why else couldn't you find this critical piece of information that would obviously be the first thing you'd want to know when you go to NASA.gov? But, it turned out, NASA had no plans for Mars. In fact, they had a crazy policy that didn't even let them talk about sending humans to Mars. Maybe, I thought, what was needed was a philanthropically funded mission to Mars to galvanize the world's attention. So I partnered with Adeo and we came up with two ideas. The first was to send a mouse on a one-way trip around Mars."

Since it was too cruel to send a mouse on a suicide mission to Mars, they pursued their second idea: sending a greenhouse instead. The Mars Oasis project was born. The project's goal was to help increase NASA's budget for Mars exploration by galvanizing public interest. Their plan was straightforward: Send a small greenhouse, stocked with seeds and dehydrated nutrient gel, to the surface of the red planet. After landing, the gel would rehydrate the seeds, the seeds would germinate, and—by seeing a photograph of the resulting plants—the world would be spurred into action. "Imagine the money shot," says Musk. "Green plants against a red Martian background."

But when Musk started looking into buying a ride to Mars, he quickly learned that launch technology had gone downhill since Apollo—that was the beginning of his epiphany. "It changed the whole plan," said Musk. "Unless something was done to reverse this degradation, a greenhouse on Mars wouldn't matter."

So why not try to reverse the degradation? Sure, space was the domain of big governments, but banking had been the domain of legacy financial institutions. So in 2002, Musk founded SpaceX, which first developed the Falcon 1 launcher, then the much more powerful Falcon 9 rocket, and the reusable Dragon capsule.[6] In May 2012, the SpaceX Dragon vehicle docked with the ISS, making history as the first commercial company to launch and dock a vehicle with the International Space Station.[7] While that greenhouse still hasn't made it to Mars, Musk recently announced that within the next fifteen years he believes he'll be able to send humans on a red planet round-trip mission for about $500,000 per person.[8]

And this is one of the first things one learns from Musk's example—he is relentless in his pursuit of the bold and, the bigger point, totally unfazed by scale. When he couldn't get a job, he started a company. When Internet commerce stalled, he reinvented banking. When he couldn't find decent launch services for his Martian greenhouse, he went into the rocket business. And as a kicker, because he never lost interest in the problem of energy, he started both an electric car and a solar energy company. It is also worth pointing out that Tesla is the first successful car company started in America in five decades and that SolarCity has become one of the nation's largest residential solar providers.[9] All told, in slightly less than a dozen years, Musk's appetite for bold has created an empire worth about $30 billion.[10]

So what's his secret? Musk has a few, but none are more important to him than passion and purpose. "I didn't go into the rocket business, the car business, or the solar business thinking this is a great opportunity. I just thought, in order to make a difference, something needed to be done. I wanted to have an impact. I wanted to create something substantially better than what came before."

Musk, like every entrepreneur in this chapter, is driven by passion and purpose. Why? Passion and purpose scale—always have, always will. Every movement, every revolution, is proof of this fact. Plus, doing anything big and bold is difficult, and at two in the morning for the fifth night in a row, when you need to keep going, you're only going to fuel yourself from deep within. You're not going to push ahead when it's someone else's mission. It needs to be yours.

But having passion and purpose is merely the first step. "The usual life cycle of starting a company begins with a lot of optimism and enthusiasm," says Musk. "This lasts for about six months, and then reality sets in. That's when you learn a lot of your assumptions were false, and that the finish line is much farther away than you thought. It's during this period that most companies die rather than scale up."

This is also where Musk urges direct and blunt feedback from close friends. "It's not going to be easy, but it's really important to solicit negative feedback from friends. In particular, feedback that helps you recognize as fast as possible what you're doing wrong and adjust course. That's usually what people don't do. They don't adjust course fast enough and adapt to the reality of the situation."

To adapt to the reality of scale, meanwhile, Musk employs a number of other strategies. We'll start with first principles, which is a lesson he borrowed from physics. "Physics training is a good framework for reasoning," explains Musk. "It forces you to boil things down to their most fundamental truths and then connect those truths in a way that lets you understand reality. This gives you a way to attack the counter-intuitive, a way of figuring out things that aren't obvious. When you're trying to create a new product or service, I think it's critical to use this framework for reasoning. It takes a lot of mental energy, but it's still the right way to do it."

In describing how this all plays out, in a 2012 interview with Kevin Rose's *Foundation*,[11] Musk talked about how first principles gave him a huge edge when developing new batteries, a key component for both Tesla and SolarCity. "So, first principles . . . What are the material constituents of the batteries? What is the spot market value of the material constituents? It has carbon, nickel, aluminum, and some polymers for separation, and a steel can. [But] if we bought that on a London metal exchange, what would each of these things cost? Oh geez . . . It's $80 per kilowatt-hour. Clearly, you need to think of clever ways to take those materials and combine them into the shape of a battery cell, but [by relying on first principles] you can have batteries that are much cheaper than anyone realizes."

It should be pointed out that first principle thinking works so well because it gives us a proven strategy for editing out complexity, while also allowing entrepreneurs to sidestep the tide of popular opinion. "[People] will do things because others are doing them," Musk explains, "because there is a trend, because they see everyone moving in one direction and decide that's the best direction to go. Sometimes this is correct, but sometimes this will take you right off a cliff. Thinking in first principles protects you from these errors."

When it comes to scale, these aren't the only errors one must guard against. Daniel Kahneman and Amos Tversky won the Nobel Prize for their work on human irrationality. One great example of this is what happens when two of the most common cognitive biases—loss aversion and narrow framing—begin to overlap. Loss aversion is the idea that humans are more sensitive to losses—even small losses—than gains, while narrow framing is our tendency to treat every risk we encounter as an isolated incident. In combination, what this means is when we go to assess risk, we tend not to look at the entire picture. In an interview with *Big Think*, Kahneman explained it like this:[12]

> People tend to frame things very narrowly. They take a narrow view of decision making. They look at the problem at hand and they deal with it as if it were the only problem. Very frequently, it's a better idea to look at problems as they will recur throughout your life and then you look at the policy that you're to adopt for a class of problems—difficult to do; would be a better thing. People frame things narrowly in the sense, for example, that they will save and borrow at the same time instead of somehow treating their whole portfolio of assets as one thing. If people were able to take a broader view, they would, in general, make better decisions.

Musk, like all the billionaires in this section, fights back. He consistently strives to broaden his view by thinking in probabilities. "Outcomes are usually not deterministic," he says, "they're probabilistic. But we don't think that way. The popular definition of insanity—doing the same thing over and over and expecting a different result—that's only true in a highly deterministic situation. If you have a probabilistic situation, which most situations are, then if you do the same thing twice, it can be quite reasonable to expect a different result."

This difference is key. Thinking in probabilities—this business has a 60 percent chance of success—rather than deterministically—if I do A and B, then C will definitely happen—doesn't just guard against oversimplification; it further protects against the brain's inherent laziness. The brain is an energy hog (it's 2 percent of our mass yet uses 25 percent of our energy), so it's always trying to conserve. As it's way more energy efficient to think in black and white, we often do. But outcomes exist across a range. "The future is not certain," continues Musk. "It's really a set of branching probability streams."

How Musk chooses which streams to explore depends on the relationship between those probabilities and the importance of his objective. "Even if the probability for success is fairly low, if the objective is really important, it's still worth doing. Conversely, if the objective is less important, then the probability needs to be much greater. How I decide which projects to take on depends on probability multiplied by the importance of the objective."

SpaceX and Tesla are great examples. When Musk started both companies, he thought their probability of success was less than 50 percent—probably a fair bit less than 50 percent—"but," he says, "I also thought these were things that needed to get done. So even if the money was lost, it was still worth trying."

Passion, probabilities, and first principles aren't just watchwords for Musk. He also backs up his words with deeds: "Between 2007 and 2009, I was in a world of hurt. Everything was going wrong. In 2008, we had the third sequential failure of the Falcon 1 rocket, Tesla couldn't raise financing because of the financial market meltdown, and Morgan Stanley couldn't honor the deal they had with SolarCity, since they were running out of money as well. There was a time when it looked like all three companies could fail. Then, on top of all of that, I was going through a divorce. That sucked. I spent my last dollar saving Tesla in 2008, and I actually went negative. I had to borrow money to pay rent."

Things turned around for Musk in late 2008. The fourth launch of Falcon 1 worked, the financial markets rebounded, SpaceX won a $1.6 billion NASA contract. And is there a lesson here as well? "The lesson I would pass on to others," he says, "the one rule I would have for entrepreneurs is, Don't leave any dollars in reserve, you can always feed yourself, but don't leave money on the table. I spent it all."

And to what result?

As TED founder Chris Anderson recently wrote in *Fortune* magazine:[13] "When you look at the incredible range of [Musk's] endeavors and search for recent comparisons in the business world, only one emerges: Steve Jobs. Most business innovations involve only incremental improvement. And of those entrepreneurs lucky enough to succeed with bigger ideas, the large majority then stick to their industry sector for expansion and consolidation. Jobs and Musk are in a category all their own: serial disrupters."

Sir Richard Branson

Just about everything Sir Richard Branson does is bold—that's his brand—so of course I wanted to interview him for this book.[14] That interview took place on a sunny September morning in 2013. We had a late breakfast at the Sunset Marquis, one of those hip Los Angeles hotels filled with celebrity sightings, then headed to the Van Nuys Airport, where I took Richard on his first zero-g flight—only one of the many adventures this global icon has racked up.

Those adventures have been captured in several biographies and countless interviews, but a few basics are worth recounting. Born on July 18, 1950, in Surrey, England, Branson struggled with dyslexia, nearly failed out of school, then dropped out at sixteen to start a youth-culture magazine called *Student*. Run by students, for students, the publication was designed, as Richard says, "to give a voice to people like me who wanted to protest against the Vietnam War and the establishment."

A rebel from the get-go and completely undaunted by scale, Branson expanded the magazine nationally and then went looking for his next opportunity. It didn't take him long to find it. As he was living in a London commune, surrounded by the British music scene, he couldn't help but notice that record stores were seriously overpriced. So he started a mail-order record company called Virgin.

The company performed modestly, but gave him enough capital to build a record shop and a recording studio. Expansion came next. Virgin signed a bevy of big acts—the Sex Pistols, Culture Club, the Rolling Stones (just to name a few)—and went on an epic ten-year run that ended with Virgin Music being one of the biggest record companies in the world. At which point, of course, Branson saw his next opportunity.

Running a music company required a considerable amount of flying, and Branson had long been frustrated by the terrible quality of the airlines. "Why, I kept wondering, couldn't we create an airline that when you walk on you feel, 'Wow, this is great.'"

Much to the dismay of his Virgin Music colleagues, that frustration launched Virgin Atlantic. "When we started," he says, "we had one used 747 and one very successful record company. Everybody at the record company was horrified by what I was doing."

And what he was doing was not easy. Branson's battle with British Airways has become the stuff of legend. At one point, in order to save his airline and avoid bankruptcy, he was forced to sell off his majority stake in Virgin Music, netting him the $800 million he needed to keep himself and his airline afloat.[15]

Despite such obstacles, Richard would build on his music business and his airline business, going on to start, invest in, and create over five hundred different companies. He founded a global empire, diversifying into everything from mobile telecommunications to trains to undersea exploration, wine distribution, fitness centers, health care clinics and, in Virgin Galactic, commercial space flight. According to the *Forbes* 2012 list of billionaires, Branson's personal worth is roughly $4.6 billion.[16] All in all, not bad for the guy who brought us *Tubular Bells*.

And if you're wondering how Branson got from *Student* to *Tubular Bells* to commercial space flight? "We're an unusual company," he says. "We're a 'way-of-life' brand—but if we weren't a way-of-life brand, we wouldn't be here today. Our first business was music stores. Music stores are dead today. But because we're about a way of life, we experimented and moved into airlines, mobile phones, and a lot of other areas. As a result, we were forced to sell our music stores—and we're alive today because of it. But if you look back at the headlines, almost every time we moved from one sector into another, the press would always say: 'Is this one step too far? Will Branson's balloon burst this time?' "

Thus the question: Why hasn't Branson's balloon burst?

Branson is a fun junkie. He has set world records in balloons. He has set world records in speedboats. He has set world records in outlandishness. Exhibit A: When Virgin Atlantic archrival British Airways decided to back the erection of a 440-foot Ferris wheel in the heart of London and had construction delays, Branson wasted no time in flying an airship over the site trailing a giant banner that read: "BA can't get it up."[17]

But what's often lost in this discussion is that fun-junkie-dom has helped Branson in two critical ways. For starters: he's immensely passionate about everything he does. When he first told Virgin Music CEOs of his idea to use one-third of last year's profits to start Virgin Atlantic, his justification was that the risk was worth it because it was "fun." "They weren't happy with the word *fun*," Branson recounted in his appropriately titled quasi-business/biography/philosophy book, *Screw It, Let's Do It*.[18] "To them, business was serious. It is. But to me, having fun matters more."

Fun matters more because Branson employs it as strategy for thinking at scale—both as a fuel (i.e., a way of harnessing his passion) and as a first principle, assuming that if something is fun for him—like an airline that makes you say "Wow!"—then it'll also be fun for everyone else. And just to make sure he's right (also because it's fun), Branson always conducts the experiment.

This is the key point. Branson's balloon hasn't burst because his fiery devotion to fun translates directly to his dedicated clientele and fervent fans. It's become a business strategy based on experimental customer-centrism. If Branson thinks a particular service might be beneficial to his customers (i.e., fun), he tries it out. This is why Virgin Atlantic was the first airline to offer free seat-back TVs, onboard massages, an onboard cocktail lounge, a glass-bottomed plane, and most recently, stand-up comedians (for now, on domestic British flights). "Unless you're customer-centric," explains Branson, "you might be able to create something wonderful, but you're not going to survive. It's about getting every little detail right. It is running your airline like you would an upscale restaurant—the kind where the owner is there every day. Virgin Atlantic started out with one plane against British Airways's hundred planes. On paper, we should not have survived. But because we were customer-centric, people went out of their way to fly us. We have survived for thirty years, during which almost every single airline that we were competing against—Pan Am, TWA, Air Florida, People's Express, Laker Airways, British Caledonian, and about twenty others—went bust."

This methodology has allowed Branson to scale. By putting his customer's needs first, Branson can triangulate vast distances, find industries that are stuck or broken, and apply his brand and experimentalism to take his shot.

But Branson, like Larry Page and Jeff Bezos, also runs his empire like a competitive ecosystem—letting some companies live, letting others die, and always, ceaselessly, experimenting. He is quick to rapidly iterate his ideas, and quicker to shut down a failure. In total, while Branson is known to have started some five hundred companies, he has also shut down the two hundred of them that didn't work.

He also gets that risk mitigation is critical. "Superficially," he says,

"I think it looks like entrepreneurs have a high tolerance for risk. But, having said that, one of the most important phrases in my life is 'protect the downside.' It should be one of the most important phrases in any businessperson's life. So okay, we made a big, bold move going into the airline business. But the most important negotiation with Boeing was that we had the right to give the plane back after twelve months. That meant I could put my toe in the water, I could see whether people liked the airline. But if it didn't work out, it wasn't going to bring everything else crashing down. I'd be able to look my record company bosses in the eye and we'd still be friends because they'd still have jobs. Protecting the downside is critical. Make bold moves but make sure to have a way out if things go wrong."

You have probably noticed by now that Branson and Musk employ different risk management strategies. In fact, all four men in this chapter have different strategies. So far we have seen Musk argue that if the idea is important enough, enormous risks are always justified. Branson also bets big, but because he's risking his entire brand (Virgin) versus a singular company (Tesla), he manages to do this is a way that doesn't jeopardize the empire.

Virgin Galactic is a fantastic example. In October 2004, when Burt Rutan demonstrated the success of the three-passenger SpaceShipOne vehicle, winning the Ansari XPRIZE, Branson and his team came in with a multi-hundred-million-dollar commitment to scale that design up to an eight-passenger vehicle able to make multiple flights per day and carry thousands into space per year. But, as is Branson's style, in 2009, he was brilliantly able to offset that risk by bringing in Aabar, the Mideast investment fund, to purchase 32 percent of Virgin Galactic for $280 million.[19] Then, two years later, Aabar increased their stake by 6 percent, committing an additional $110 million to fund small satellite launch capability.[20] So, sure, Branson bet a huge amount on Virgin Galactic, but he then protected that investment and brought in an extra $390 million in working capital to ensure its success. Branson, it seems, isn't just bold in his risk taking, he's also bold about his risk mitigation. The end result, though, is the same.

A few years back Google's April Fool's joke was the announcement of a new company called Virgle, a fake Google/Virgin collaboration to establish a permanent human settlement on Mars.[21] There were corresponding videos on YouTube, with Branson, Brin, and Page talking about where to fill out a colonist application and how they were currently searching for experts in physics, engineering, and—most critically—Guitar Hero III. But the really funny part is that a lot of people didn't realize it was a joke. A lot of others still aren't sure. Which is to say, Branson's appetite for bold is so big and his track record at scale so stellar that, for a great many, it's difficult to *not* believe Branson is going to Mars.

Jeff Bezos

Jeff Bezos is a busy man. About five years ago, when I emailed him to set up a breakfast meeting, his response came back: "Peter, I'm so busy I'm trying to optimize my toothbrushing time." And there's a reason he's so harried—the same reason Eric Schmidt listed Amazon (alongside Google, Apple, and Facebook) as one of the four horsemen of technology. Bezos isn't interested in small shifts or polite progress. He wants to effect change on a massive scale, with customer-centric thinking and long-term thinking being the primary drivers behind this revolution.

Jeff Bezos was born on January 12, 1964, in Albuquerque, New Mexico.[22] Like Musk, he too showed an early interest in how things work. As a toddler, he disassembled his crib with a screwdriver. Later he rigged up a series of elaborate electric alarms to keep his siblings out of his room. His childhood in Houston was a wash of science projects, science fairs, and *Star Trek* episodes. His high school years were spent in Miami, which is also where his love of computers arose. As his brother, Mark Bezos, once told reporters:[23] "He would certainly have been classified as 'the nerd.'"

Defending that title, Bezos went to college at Princeton, where he finished summa cum laude in 1986 with a degree in computer science and electrical engineering. After graduation, Bezos pursued investment banking, becoming the youngest vice president at the Wall Street firm of D. E. Shaw. But he was not destined for a career in finance. A short four years later, Bezos had an epiphany that caused him to quit his lucrative job, move to Seattle, and attempt to change the world—one e-commerce transaction at a time.

"The wake-up call that led to Amazon.com was finding that web usage in the spring of 1994 was growing at 2300 percent a year," said Bezos during a speech given in 2001 at the Academy of Achievement in Washington, DC.[24] "And things just do not grow that fast. . . . You could tell anecdotally, even though there wasn't good research on this at the time, that the baseline of web usage wasn't trivial . . . so the question was, What kind of business plan would make sense in the context of that growth? And I went through a whole bunch of different things. I made a list of twenty different products, looking for the first product to sell online. I came up with books for a bunch of reasons, but primarily because books were very unusual in one respect . . . there are more of them than there are products of any other category. So there are literally millions of different books in print . . . and computers are good at organizing such large selections of products. And you could build something online that literally couldn't be built in any other way. You couldn't have a physical world bookstore or a paper catalog with millions of different books."

In the early days, Amazon's success was by no means a given, but Bezos has always been a fantastic evangelist. Also an honest evangelist. When his parents decided to invest a good portion of their life savings in the company, Jeff—in a great example of probabilistic thinking—told them they had a "70 percent chance of losing their money." He also admitted to hedging his bet. "I was giving myself triple the normal odds, because, if you look at the odds of a start-up succeeding at all, it's only about 10 percent. Here I was giving myself a 30 percent chance."

Bezos used his parents' money to set up shop in the proverbial garage of his Seattle home, soon expanding into a nearby two-bedroom house. It was from there, on July 16, 1995, that Amazon.com opened for business. Bezos and his small team designed a small launch. They invited a couple of hundred friends and family to visit the site, and were so excited about potential business, they hooked up an electronic bell to ring every time a transaction occurred. "There was a time when we were examining every order that came in," says Bezos, "and it was always a family member placing the order. [But] the first order we got from a stranger—I remember there were probably ten of us in the company, all gathered around after the bell rang, looking at the order. We were like: 'Is that your mom?' 'That's not my mom!' And thus it began."

And did it ever.

The bell was soon ringing continuously (they had to disconnect it). Within one month, Amazon had customers in forty-five countries and all fifty U.S. states. Within two months, sales had reached $20,000 a week. Then, in May of 1997, they went public with a $500 million valuation. Six months after that, the number climbed to $1.2 billion, then rose to $23 billion over the next two years. Bezos, now thirty-five years old, had gone from "I have a neat idea" to "I run a multibillion-dollar company" in just over five years.[25]

Bezos's success sits atop two critical strategies: long-term thinking and customer-centric thinking. We'll take them one at a time.

Bezos has never been interested in quick profits or short-term rewards. From the start, Amazon has been playing the long game. In his now-famous 1997 letter to his shareholders,[26] Bezos put it this way: "We believe that a fundamental measure of our success will be the shareholder value we create over the *long term*. . . . Because of our emphasis on the long term, we may make decisions and weigh trade-offs differently than some companies."

The letter went on to explain his thinking strategy with a number of now-famous points:

- We will continue to make investment decisions in light of long-term market leadership considerations rather than short-term profitability considerations or short-term Wall Street reactions.

- We will make bold rather than timid investment decisions where we see a sufficient probability of gaining market leadership advantages. Some of these investments will pay off, others will not, and we will have learned another valuable lesson in either case.
- We will share our strategic thought processes with you when we make bold choices (to the extent competitive pressures allow), so that you may evaluate for yourselves whether we are making rational long-term leadership investments.
- We will balance our focus on growth with emphasis on long-term profitability and capital management. At this stage, we choose to prioritize growth because we believe that scale is central to achieving the potential of our business model.

This letter is often held up as the encapsulation of Bezos's view on the subject, but personally, I think an answer he gave to an Amazon Web Services Live audience in 2012[27] was far more revealing:

"What's going to change in the next ten years?" And that is a very interesting question; it's a very common one. I almost never get the question: "What's not going to change in the next ten years?" And I submit to you that that second question is actually the more important of the two—because you can build a business strategy around the things that are stable in time. . . . In our retail business, we know that customers want low prices, and I know that's going to be true ten years from now. They want fast delivery; they want vast selection. It's impossible to imagine a future ten years from now where a customer comes up and says, "Jeff, I love Amazon; I just wish the prices were a little higher" [or] "I love Amazon; I just wish you'd deliver a little more slowly." Impossible. And so the effort we put into those things, spinning those things up, we know the energy we put into it today will still be paying off dividends for our customers ten years from now. When you have something that you know is true, even over the long term, you can afford to put a lot of energy into it.

It's also in the above passage that Bezos raises the second secret to his success: radical customer-centrism. This too has been there since the beginning. To return to his 1997 letter to shareholders, we can find the idea encapsulated in one of his bullet points: "we will continue to focus relentlessly on our customers," and then reinforced at the letter's close:

> From the beginning, our focus has been on offering our customers compelling value. We realized that the Web was, and still is, the World Wide Wait. Therefore, we set out to offer customers something they simply could not get any other way, and began serving them with books. We brought them much more selection than was possible in a physical store (our store would now occupy six football fields), and presented it in a useful, easy-to-search, and easy-to-browse format in a store open 365 days a year, 24 hours a day. We maintained a dogged focus on improving the shopping experience, and in 1997 substantially enhanced our store. We now offer customers gift certificates, 1-Click shopping, and vastly more reviews, content, browsing options, and recommendation features. We dramatically lowered prices, further increasing customer value. Word of mouth remains the most powerful customer acquisition tool we have, and we are grateful for the trust our customers have placed in us. Repeat purchases and word of mouth have combined to make Amazon.com the market leader in online bookselling.

But it's the combination of long-term thinking and customer-centrism that has helped Amazon extend their reach far beyond books. Bezos has ventured into music, movies, toys, electronics, automotive parts, and well, just about everything. They have also continued to surround their original market, moving from books into ebooks and ebook readers (with Kindle), and most recently, publishing itself. Meanwhile, Amazon Web Services—their cloud business—has become a beast in its own right (worth nearly $3 billion, according to a November 2013 *Business Insider* analysis).[28] As Morgan Stanley analyst Scott Devitt told the *New York Times*:[29] "Amazon is marching to a different drumbeat, which is long term. Are they doing the right thing? Absolutely. Amazon is growing at twice the rate of e-commerce as a whole, which is growing five times faster than retail over all. Amazon is bypassing margins and profits for growth."

Bezos also understands that the only way to really succeed with his long-term customer-centrism is via experimentation. He also knows that this approach will occasionally produce spectacular failure. As he recently said to a group at the Utah Technology Council Hall of Fame[30] dinner: "The way I think about it, if you want to invent, if you want to do any innovation, anything new, you're going to have failures because you need to experiment. I think the amount of useful invention you do is directly proportional to the number of experiments you can run per week per month per year. So if you're going to increase the number of experiments, you're also going to increase the number of failures.

"And if you're going to invent, you've got to be willing to be misunderstood for long periods of time. Anything new and different is initially going to be misunderstood. It will be misunderstood by well-meaning critics, who are worried that it might not work out. It will be misunderstood by self-interested critics, who have a profit stream connected to the old way. Either way, if you can't weather this kind of misunderstanding and criticism, then whatever you do, don't do anything new."

Bezos, though, can't get enough of the new. His sideline aerospace business, Blue Origin,[31] is trying to solve the long-standing rocketry puzzle of vertical takeoff and landing, and, in a sort of an aerospace-meets-Amazon cross-pollination, in December of 2013, Bezos announced that some time in the next five years, drones will be delivering packages for Amazon,[32] enabling both half-hour delivery and Bezos's big dream of getting into the food market and thus finally being able to unseat Walmart from its throne position and turning Amazon into the most successful "Everything Store" in history.

Bezos discussed his drones in his 2014 shareholder letter,[33] hitting again on his theme of experimentation and rapid iteration: "Failure comes part and parcel with invention. It's not optional. . . . We understand that and believe in failing early and iterating until we get it right. When this process works, it means our failures are relatively small in size (most experiments can start small), and when we hit on something that is really working for customers, we double down on it with hopes to turn it into an even bigger success."

At TED 2014, when I asked Bezos what advice he would give to exponential entrepreneurs, he, like Musk, counsels a focusing on passion, not fads.

"It's so hard to catch something that everybody already knows is hot," says Bezos. "Instead, position yourself and wait for the wave to come to you. So then you ask, Position myself where? Position yourself with something that captures your curiosity, something that you're missionary about. I tell people that when we acquire companies, I'm always trying to figure out: Is this person who leads this company a missionary or a mercenary? The missionary is building the product and building the service because they love the customer, because they love the product, because they love the service. The mercenary is building the product or service so that they can flip the company and make money. One of the great paradoxes is that the missionaries end up making more money than the mercenaries anyway. And so pick something that you are passionate about, that's my number one piece of advice."

Larry Page

In November 2004, about a month after the Ansari XPRIZE was won, I found myself at the Googleplex, making a presentation to a couple of thousand Googlers on the future of space travel. Afterwards, I was flooded by people wanting to continue the discussion. Last in line was a young guy in his early thirties, wearing a black T-shirt and carrying a backpack. "Hi," he said. "I'm Larry Page, want to have lunch?"

During lunch, we covered everything from robots to space elevators to autonomous cars. Immediately clear was Larry's intense curiosity about all things technical and his insatiable desire to push boundaries. His favorite questions were "Why not?" and "Why not bigger?" Clearly I was dealing with someone unaccustomed to limits. That conversation took place almost a decade back and little has changed, except the fact that, well, everything has changed.

Born March 26, 1973, in East Lansing, Michigan, Page had a hereditary predilection for computers.[34] His mother, Gloria, was a computer science professor at Michigan State; his father, Carl, was a pioneer in both computer science and artificial intelligence. Not surprisingly, Larry became the first kid in his elementary school to turn in an assignment from a word processor.

Page went on to get a degree in computer engineering from the University of Michigan—where he famously created an inkjet printer made of LEGO bricks—then went off to Stanford for a PhD in computer science. While searching for a theme for his dissertation, Page got curious about the mathematical properties of the web, specifically the idea that its link structure was based on citations and that these citations could be represented as a huge graph. This led to a partnership with another Stanford PhD student, Sergey Brin, and a research project nicknamed BackRub, which led to the page-rank algorithm that became Google. Not surprisingly, neither Brin nor Page ever finished their PhDs.

Instead, in 1998, they dropped out and started up and changed history. The PageRank algorithm democratized access to information, or as a recent article in *Wired* put it: "Search, Google's core product, is itself wondrous. Unlike shiny new gadgets, however, Google search has become such an expected part of the internet's fabric that it has become mundane."[35] Meanwhile, YouTube became the dominant video platform on the web, Chrome the most popular browser, and Android the most prolific mobile phone operating system ever. To put this in perspective, today a Masai warrior in the heart of Kenya who has a smartphone and access to Google has—at his fingertips—access to the same level of information that the president of the United States did eighteen years ago.

And it's this kind of world-changing impact that especially interests Page. In an impromptu speech given at the Singularity University founding conference, Larry stood up in front of an audience of some 150 attendees and said: "I have a very simple metric I use: Are you working on something that can change the world? Yes or no? The answer for 99.99999 percent of people is no. I think we need to be training people on how to change the world."

Today Larry's desire to make good on this promise and Google's vision for the future have become indistinguishable, and his actions have pushed the company to new heights and riches. In the three years following Page's promotion from copresident to CEO in 2011, the company's value has doubled to $350 billion (with Page's 16 percent worth about $50 billion), its cash war chest has risen to $75 billion, and its annual research and development budget increased to $8.5 billion.[36] And Page, as visionary CEO, can spend that money almost anywhere he pleases. Autonomous cars, augmented reality, ending aging, ubiquitous Internet—clearly, what pleases Page is the big and bold.

In 2012, I presented at Google Zeitgeist, their annual customer conference. The organizers had slotted me at the end of the second day, asking me to give an uplifting speech with my *Abundance* message. Afterward, Page followed me onstage to deliver closing remarks, which was when I learned the origin of his appetite for bold. "When I was a student at the University of Michigan," he said, "I took this summer leadership course. Their slogan was: 'Have a healthy disregard for the impossible.' That's stuck with me all of these years. I know it sounds kind of nuts, but it's often easier to make progress when you're really ambitious. Since no one else is willing to try those things, you don't have any competition. And you get all the best people, because the best people want to work on the most ambitious things. For this reason, I've come to believe that anything you imagine is probably doable. You just have to imagine it and work on it."[37]

What helps Page imagine the impossible is a fervent belief in rational optimism.[38] The term, borrowed from author Matt Ridley, refers to the exact kind of optimism we advocated for in *Abundance*. It does not mean pie-in-the-sky daydreaming. It means rather a sober review of the facts, which include the fact that technology is accelerating exponentially and transforming scarcity into abundance, that the tools of tomorrow are giving us ever-increasing problem-solving leverage, that the world—based on dozens of metrics (see the *Abundance* appendix)—is also getting exponentially better, and finally, as a result, that small teams are now more empowered to solve grand challenges than ever before. And it's these reasons that make rational optimism such an important strategy for thinking at scale. Or as Larry Page famously said in his I/O 2013 keynote: "Being negative is not how we make progress."[39]

This is not a passing sentiment for the Google chief, it's a core philosophy. "I'm tremendously optimistic," says Page. "I'm certain that whatever challenges we take on, we can solve with a little bit of concerted effort and some good technology. And that's an exciting place to be. [It means] our job is really to make the world better. We need more people working on this. We need to have more ambitious goals. The world has enough resources to provide a good quality of life for everyone. We have enough raw materials. We need to get better organized and move a lot faster."

Speed is of the essence for Page, which is why he backs up his rationally optimistic view of the future with an extraordinarily healthy appetite for risky, bold innovation. He has become famous for pushing people far beyond their comfort zones. When he recruited Sebastian Thrun to develop their autonomous car, Page declared that 100,000 miles was the incredible target for how far that car was supposed to be able to drive on its own (today the car has driven well over 500,000 miles).[40] When Google wanted to do simultaneous translation between languages, they found some machine-learning researchers and, as Page explains, "We asked them, 'Do you think you can set up an algorithm to translate between any two languages and do it better than a human translator?' They laughed at us and said it was impossible. But they were willing to try. . . . And now, six years later we can translate between sixty-four different languages. In many languages, we're better than an average human translator and we can do it instantly and for free."[41]

Or, to offer an even more colorful example, in a Steven Levy story for *Wired*, Astro Teller talked about wheeling an imaginary time machine into Page's office, plugging it in, and then demonstrating that it works. "Instead of being bowled over," says Teller, "Page asks why it needs a plug. Wouldn't it be better if it didn't use power at all? It's not because he's not excited about time machines or ungrateful that we built it. It's just core to who he is. There's always more to do, and his focus is on where the next 10x will come from.' "[42]

So where does the next 10x come from?

For Page, like Musk and Branson and Bezos, that answer always sits at the intersection of long-term thinking and customer-centric thinking:

> We always try to concentrate on the long term. Many of the things we started—like Chrome—were seen as crazy when we launched them. So how do we decide what to do? How do we decide what's really important to work on? I like to call it the "toothbrush test." The toothbrush test is simple: Do you use it as often as you use your toothbrush? For most people, I guess that's twice a day. I think we really want things like that. We use Gmail much more than twice a day. And YouTube. Those things are amazing. Yet, when we first looked at YouTube, people said, "Oh, you guys are never going to make money with that, but you bought it for $1.4 billion. You're totally crazy." And, you know, we were reasonably crazy, but it was a good bet. We've actually been doubling revenue on YouTube every year for four years. And if you're doubling things, no matter where you start from, it starts to add up pretty quickly. Our philosophy is that the things that people use often are really important to them and we think that over time, you can make money from those things.[43]

It's also for these same reasons that Page has devoted considerable resources to the pursuit of AI. "Artificial intelligence would be the ultimate version of Google," he explains. "The ultimate search engine. It would understand everything on the web, it would understand exactly what you wanted, and it would give you the right thing. And we're nowhere near doing that now. However, we can get incrementally closer to doing that. And that's what we're working on."

In the spring of 2014, when Page hosted a group of XPRIZE donors at GoogleX, he reflected on the benefits of bold ambitions. "You'd think that as we do more ambitious things, our failure rates would go up, but it doesn't really seem to. The reason, I believe, is even if you fail in doing something ambitious, you usually succeed in doing something important. I like to use the example of our first attempt at creating AI, which was started when Google had less than two hundred people back in 2000. We didn't succeed in creating an AI, but we did come up with AdSense, where we target search ads against web pages, which has become a good chunk of our revenue. So we failed at making AI, but we got distracted by something useful. Pretty much 100 percent of these things have gone that way."

Of course, AI is not the only futuristic technology they're working on. Google is also the title sponsor for the $30 million Google Lunar XPRIZE—with the goal being to put the first robot on the Moon (seen as a first step toward extending humanity's reach and economic influence beyond the Earth), which is to say that, just like Bezos, Branson, and Musk, Page too dreams of outer space. But his ambitions don't end there. Unlike the others, Page has taken an even more adventurous step, announcing the longevity start-up Calico in November of 2013,[44] which is Google's entry into the anti-aging world, or, as *Time* magazine put it: "Google vs. Death."[45]

In his widely circulated essay "Google Wins Everything," Internet entrepreneur and blogger Jason Calacanis puts it this way: "If you work for Larry and are not thinking 10x, don't expect to keep your job for very long. That insanity-by-design is creating a one-upmanship that hasn't been seen in the history of mankind. Larry's campaign, if successful, will make Caesar, Napoleon, Columbus, the Wright Brothers, the Apollo 11 Mission, the Manhattan Project, and our Founding Fathers look limited in scope."[46]

PART THREE

THE BOLD CROWD

CHAPTER SEVEN

Crowdsourcing

Marketplace of the Rising Billion

It's the fall of 2000. There are now more than 20 million websites on the Internet.[1] The browser wars (AOL versus Netscape) are in full swing. And with the recent bursting of the dot-com bubble, there are a lot of out-of-work graphic designers hanging around cyberspace, just looking for something to do.

Jake Nickell and Jacob DeHart are among them.

Nickell and DeHart are both nineteen years old. They too are out-of-work designers. They met during an online T-shirt design competition—something that was then occasionally starting to happen—and decided they wanted such contests to happen more frequently. Instead of a competition just once a year, they decided to create a website that hosted them once a week. Anyone with a good T-shirt design could enter. Everyone in the community could vote. The winner got a hundred dollars, and the winning T-shirt was put up for sale on the site. They named their new venture Threadless.com, and mostly it seemed harmless enough.[2]

Turns out, people liked to vote on T-shirts. They *really* liked to vote on T-shirts. Within a few years, Threadless was turning an annual profit north of $20 million. Almost unintentionally, Nickell and DeHart had become the third largest T-shirt manufacturers in the United States.

And Threadless wasn't alone in finding ways to tap into the burgeoning online community. During this same period, a software designer named Philip Rosedale noticed that hardcore gamers weren't just interested in playing games; they also seemed to want to spend their time designing the games themselves. So he created Second Life, a massive virtual world that was essentially built for free, with Rosedale merely outsourcing software development to the gaming crowd. And the crowd, as Jeff Howe wrote in *Wired*, "[was] only too eager to do the work."[3]

So eager, in fact, that throughout the early 2000s, the Second Life community generated 10,000 developer-hours worth of content a day. An entire economy emerged inside the game. Right around the time that Threadless was starting to make $20 million in annual profits, *Business World* put Anshe Chung on their cover—the very first virtual citizen who had become a real-life millionaire because of his Second Life business.[4]

It was also Jeff Howe, alongside *Wired* editor Mark Robinson, who noticed what was happening with the likes of Threadless and Second Life and coined the term *crowdsourcing*. Howe defined the word as "the act of a company or institution taking a function once performed by employees and outsourcing it to an undefined (and generally large) network of people in the form of an open call. This can take the form of peer-production (when the job is performed collaboratively), but is also often undertaken by sole individuals. The crucial prerequisite is the use of the open call format and the large network of potential laborers."[5]

As crowdsourcing gained steam, crowdfunding (covered in detail in the next chapter) was developing. While the idea dates back to the 1980s, it became a mainstream phenomenon in 2005, when Kiva.org became the first microlending website, tapping the crowd to provide tiny loans (usually less than $100) to entrepreneurs in developing countries. By 2009, Kiva had distributed over $100 million in loans, with a staggering 98 percent repayment rate. By 2013 that number had jumped to $526,460,675 in loans from 1,047,653 Kiva lenders while maintaining a 98.96 percent repayment rate.[6]

This was also the same time when crowdfunding sites like Indiegogo and Kickstarter came into being, giving birth to a new way to raise money for creative projects. Want to make a movie? Cut a new CD? Design a new kind of watch? Just put a video up on either of these sites and ask the crowd for the money. It didn't take long before the *New York Times* started calling Kickstarter "the people's NEA [National Endowment for the Arts]," and well, they weren't kidding.[7] In 2010, the site raised over $27 million and funded 3,910 projects. The following year, the amount raised jumped to $99 million, funding 11,836 projects. In 2013, these figures were north of $480 million and some 19,911 projects.[8] And Kickstarter is only one example. While Indiegogo has not released its growth numbers, in early 2014, its success was impressive enough to command a $40 million equity investment, the largest venture investment yet for a crowdfunding start-up. All told, across the entire sector, hundreds of crowdfunding platforms have materialized, giving entrepreneurs access to what will soon exceed tens of billions of dollars in annual funding.

As movements, both crowdfunding and crowdsourcing diversified quickly, with all sorts of commercial applications beginning to emerge. The graphic design hub 99designs, for example, allows users to submit a design need and an associated budget—say, a new logo for $299—and the crowd competes for the business. Gengo.com offers crowdsourced human translators, CastingWords does audio transcription, and Maven Research—aka the global knowledge marketplace—provides expertise in hundreds of thousands of disciplines.

Big business has also gotten in on the action. Anheuser-Busch now relies on the crowd to craft beers. General Mills has tapped them for everything from packaging design innovation to novel ingredient suggestions. Scientific research has become another growth area. The Polymath Project pits the crowd against unsolvable math problems, Foldit harnesses them for protein folding, and Zooniverse allows anyone to categorize galaxies, discover new planets and even hunt for alien life.

So why does this whole crowdsourcing arena matter so much for exponential entrepreneurs? Consider what Larry Page's dream of artificial intelligence might look like when it finally arrives. This would be a system that understands your intentions and desires and can help turn them into reality. Make your request to JARVIS and the AI will analyze data, write programs, create, design, and—probably via 3-D printing—manufacture exactly what you require, whenever you require it. Everything from the prototyping of new products to getting real-time data-mining insights about an entire market is in the offing. Sounds exciting, right? But while that extraordinary capability may still be a decade or so away, in the interim, we have the crowd.

Part three of *Bold* examines what we've chosen to call *exponential crowd tools*—all the various permutations of crowd-powered capabilities now available to everyone. These tools are exponential in power for three simple reasons. First, over the next decade, the size of the crowd (those folks online) is expected to more than double—from roughly 2 billion to 5 billion people (perhaps 7 billion if some of the orbital or stratospheric communication solutions are deployed).[9] This means 3 billion new minds are about to join the global conversation (this is the group referred to in *Abundance* as the rising billion). Second, the communication technologies underpinning the crowd are growing exponentially, morphing once thin data connections into ubiquitous broadband. The multinational professional services firm PricewaterhouseCoopers projects that the penetration of mobile (broadband) Internet services will reach 54 percent of the world's population by the end of 2017.[10] As a result, the crowd is becoming hyperconnected and hyperresponsive. Third, and perhaps most important, all of the exponential technologies discussed in part one of this book are starting to become easily accessible to the masses, further empowering this hyperconnected, hyperresponsive crowd. What this means is that the people you can now tap for support are themselves far more capable than ever before.

Global fixed-broadband and mobile Internet penetration (%) 2008–2017

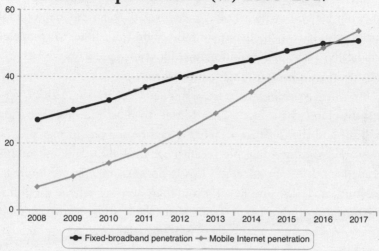

Internet Penetration: The Rising Billions

Source: http://www.pwc.com

Before we launch into crowdsourcing in greater detail, it's helpful to pull back a bit and see how these ideas work within the greater context of part three of this book. In this chapter, we'll dive deeper, coming to understand how just about anything you need done can now be done by some crowd-powered platform. We'll also come to understand why crowdsourcing works so well and, critically, how you can get the most out of it. In addition, just in case you'd like to start your own crowdsourcing site, I'll cover the challenges that platform founders had in creating their own and the lessons they learned along the way. Finally, if you are interested in leveraging existing platforms, I'll introduce you to some of the most powerful services tailored to the exponential entrepreneur.

Next, in chapter 8, we'll shift our examination to crowdfunding, one of the greatest capital-raising tools available to today's entrepreneur, providing you with an in-depth overview of the sector and a step-by-step guide to tapping its full potential. Then, in chapter 9, we'll see how to build a passionate, committed, and capable "exponential community" to support and advance your boldest ideas. Finally, chapter 10 closes out the book with a look at incentive prizes, an exponential tool that lets you harness basic human motivation (our desire to compete) to radically accelerate innovation, providing entrepreneurs with exceptional leverage when tackling grand challenges.

It's worth pausing to unpack this last idea further. Until very recently, grand challenges were off-limits to most mortals. The issue was scale and the fact that scale has always been a pay-to-play proposition. Historically, going big meant huge capital outlays and multidecade bets. It meant arms and legs in dozens, sometimes hundreds, of countries. It also meant an astounding array of talent, the infrastructure to hire that talent, retain that talent, and—as the technology evolved—retrain that talent. But with the array of exponential crowd tools available to today's entrepreneur, the entire playing field has shifted. Today, incredibly, the ability to play at scale is never more than a few mouse clicks away.

To best examine all these exponential crowd tools, throughout part three, we'll be following a format similar to the one introduced in our discussion of 3-D printing. First, we'll examine the past, present, and future: the history of the tool, its current state, and its future possibilities. Next, we'll dive into three case studies, seeing how other entrepreneurs have leveraged these exponential organizational tools to tackle the big and bold. Each chapter will close with a very concrete how-to section, intended to allow anyone to finish the reading and jump immediately into the doing.

To gather the advice in the how-to sections in part three, my team did exhaustive research, interviewing over a hundred top platform providers, the very people behind all of these crowd-powered companies, and speaking with top users, those exponential entrepreneurs who have already succeeded in leveraging crowd tools to tackle the bold. We also conducted a meta-analysis of all the various how-to articles online and in major reports, distilling key lessons and insights. Finally, during the same time this work was going on, I had the opportunity to implement much of this advice, putting it to the test in my own companies. Taken together, my hope is that these how-to sections serve as a comprehensive playbook, literally a user's guide for going big, creating wealth, and impacting the world.

Let's begin.

Case Study 1: Freelancer—Quantum Mechanic for Hire by the Hour[11]

It started back in the late 2000s. Matt Barrie was irritated. A venture capitalist and entrepreneur with expertise in information security, Barrie was coding a website and trying to hire someone—anyone—to do some basic data entry. His rates were decent. He was willing to pay two dollars a line to the kid brother or kid sister of a friend. But there was soccer practice. There were exams. The whole process dragged on for months. It wasn't working at all.

"In frustration," says Barrie, "I got online and posted the job on a site called Get a Freelancer. Three hours later, I came back to my computer and found seventy-four emails from people willing to do it for anywhere from a hundred dollars to a thousand. I hired a team in Vietnam that finished the job in three days. It was perfect. I didn't have to pay them until everything was done. The whole process was mind-blowing."

Following this revelation, Barrie began buying up existing crowdsourcing companies. He started with Get a Freelancer, the first site he'd used, then moved on to Scriptlance and vWorker and soon added seven more. All nine were merged into Freelancer.com, which quickly became a behemoth. The numbers are impressive. In less than half a dozen years, the site has grown to 10 million users, and has become the largest freelancer marketplace on the planet. Over 5.4 million jobs have been posted, representing a total value of $1.39 billion in work. There are members in 234 countries and regions around the world, with about 75 percent of the workers coming from countries such as

India, Pakistan, Bangladesh, the Philippines, and China. "It's a very, very long tail," says Barrie. "Only about 25 percent of the job postings come from companies. The vast majority come from either individuals or small businesses."

Under the hood, Freelancer.com is actually in the connection business, existing to bring together two types of entrepreneurs. "On one side," explains Barrie, "we've got under-resourced small-business entrepreneurs in the developed world. They don't have a lot of money, don't have a lot of time, but they have all these ideas. On the other side, in the developing world, we empower a whole new class of entrepreneur— the service providers who can help turn those ideas into reality."

How diverse are the experts on Freelancer.com? "There are a lot of people on the site who are moonlighting," says Barrie. "So we're not talking plumbing or pest control. We're talking any job you can imagine. We've got PhDs on the site. I've seen both quantum physics and aerospace jobs handled perfectly. From a macroscopic perspective, freelancing is really the vanguard of an economic revolution that's sweeping through the developing world as people can now wake up and [say], 'Hey, I want to work in this very niche area in technology. Maybe there's no jobs locally, but now I can work for a global client base and earn fantastic income.'"

What this means is that entrepreneurs can do much more than just outsource work via Freelancer—you can actually build whole businesses on its back. Take Barrie's business partner, Simon Clausen, who started out as one of Australia's top technology entrepreneurs. When Clausen was building his antivirus company, PC Tools, he began by crowdsourcing his first antivirus app—paying an Indian company a thousand dollars for the program. And it worked. PC Tools got to $100 million in revenue per annum before selling to Symantec.

Barrie summarizes the potential nicely, "Today you get someone to analyze the data, put together beautiful figures and graphs, crunch the numbers, do mathematical modeling. It's as sophisticated as you think. As for the future, you're only limited by your imagination." The *New York Times* columnist Thomas Friedman put it like this: "You have a spark of idea now. You can get a designer in Taiwan to design it. You can get the prototype produced in China. In Vietnam you can get it mass-produced. On Freelancer they can do your back office, your logo and so forth. I mean, really—now you can be one guy sitting in a room with a few thousand dollars and off the back of a credit card you can build a multimillion-dollar company."[12]

Case Study 2: Tongal—Genius TV Commercials at One One-Hundredth the Price[13]

L.A., where I live, is something of a company town, with Hollywood being the company. As a result, every coffee shop and bus stop is packed full of scriptwriters, producers, and directors. Mix that incredible talent density with the plummeting cost of 1080p high-definition cameras and the awesome editing software available on every Mac and you have the making of a video production revolution.

No one has done a better job of exploiting this for *your* benefit than Tongal, a crowdsourcing platform that can help you create TV-quality video commercials for digital or television advertising ten times cheaper, ten times faster, and with ten times the number of content options than by standard processes. Tongal, like Freelancer.com, was also born of frustration—in this case, the frustration of James DeJulio.

DeJulio began his career in investment banking, but quickly realized that finance was not the world for him. So he decided to try his hand at Hollywood. And, like a lot of gifted and highly qualified people, DeJulio started at the bottom. "I couldn't believe how hard it was to get a job that paid so little," he recalls.

DeJulio eventually landed a job in production with Paramount, where he rose to vice-president and was the force behind *How to Lose a Guy in 10 Days* and *The Kid Stays in the Picture*, but he soon soured—growing as disappointed in Tinseltown as he had been with finance. "I was frustrated by how many good ideas never saw the light of day, how there was a small list of people who tightly controlled all the creative work, and how many talented people there were who couldn't break into the system."

Turns out, *The Da Vinci Code* was his breaking point. DeJulio's boss had gotten a pre-publication copy of Dan Brown's soon-to-be mega-bestseller and asked DeJulio to take a look. "I read it, I thought it was a real page-turner and I gave it to my boss, saying, 'This is really exciting. The studio should make this film.'" But the studio passed it off to yet another person who decided the book had "no real entertainment value." That summer, when everybody in America was reading the thriller, DeJulio decided there had to be a better way.

That was about the time he bumped into Jack Hughes, founder of the crowdsourcing software-solutions company TopCoder, who helped him realize that the same distributed, crowd-powered approach that TopCoder employs in helping companies fulfill their software needs would work in Hollywood. "I started to think about how we could turn the industry on its head and attack the video content creation problem in a very different, incentive-based way," says DeJulio. "There were so many people around with an HD camera and a Mac who really want to do this work."

DeJulio wasn't overstating the case. Today the Tongal freelance pool includes more than 40,000 creatives, primarily those working on short-form videos, commercials, and such, and has created content for major brands such as Unilever, LEGO, Pringles, and Speed Stick. And they have done so in a way that is far faster, more creative, and more cost-effective than the traditional approach.

"A large brand will typically spend between 10 and 20 percent of their media buy on creative," DeJulio explains. "So if they have a $500 million media budget, there's somewhere between $50 to $100 million going toward creating content. For that money they'll get seven to ten pieces of content, but not right away. If you're going to spend $1 million on one piece of content, it's going to take a long time—six months, nine months, a year—to fully develop. With this budget and timeline, brands have no margin to take chances creatively."

By contrast, the Tongal process: If a brand wants to crowdsource a commercial, the first step is to put up a purse—anywhere from $50,000 to $200,000. Then, Tongal breaks the project into three phases: ideation, production, and distribution, allowing creatives with different specialties (writing, directing, animating, acting, social media promotion, and so on) to focus on what they do best. In the first competition—the ideation phase—a client creates a brief describing its objective. Tongal members read the brief and submit their best ideas in 500 characters (about three tweets). Customers then pick a small number of ideas they like and pay a small portion of the purse to these winners.

Next up is production, where directors select one of the winning concepts and submit their take. Another round of winners are selected and these folks are given the time and money to crank out their vision. But this phase is not just limited to these few winning directors. Tongal also allows anyone to submit a wild card video. Finally, sponsors select their favorite video (or videos), the winning directors get paid, and the winning videos get released to the world.

Compared to the seven to ten pieces of content the traditional process produces, Tongal competitions generate an average of 422 concepts in the idea phase, followed by an average of 20 to 100 finished video pieces in the video production phase. That is a huge return for the invested dollars and time.

And the level of talent that brands have access to via Tongal continues to grow. "In the beginning," says DeJulio, "the majority of the creative people working in the Tongal community were hobbyists who grew up making content for the Internet, but as our prize purses have steadily increased in value, we're starting to see super-talented people—people who would otherwise have been hired in traditional advertising industry—opt for our platform instead. And as Tongal-generated content gets better, brands are putting more money on the line. It's a very positive, self-reinforcing cycle. So now it's not unusual to have a $50,000 or $60,000 prize purses result in a set of deliverables for which a traditional agency would have normally charged millions."

So how good can it get? How about good enough to make it to the Super Bowl? In 2012, Tongal ran a $27,000 challenge for Colgate Palmolive's Speed Stick deodorant to create a thirty-second piece for digital (Internet) placement. But the final ad was so good, Colgate Palmolive placed it into one of its coveted Super Bowl slots. The ad actually finished at 24 (out of 60) on *USA Today*'s ad meter, well ahead of more than thirty other ads created traditionally, and with budgets literally 500 times larger. The television audience for the Super Bowl has been estimated at more than 110 million people, and on YouTube this Speed Stick commercial has been seen almost 1.2 million times. Pretty good for a $27,000 investment.

Case Study 3: reCAPTCHA and Duolingo— Dual-Use Crowdsourcing[14]

Carnegie Mellon computer scientist Luis von Ahn wasn't entirely pleased with himself. Back in 2000, Ahn was part of the team of people who invented the challenge-response test known as CAPTCHA—those squiggly, drunken characters we have to recopy to log on to certain websites. The purpose of CAPTCHA was to help differentiate bots from humans, but what was bugging Ahn was CAPTCHA's success.

"All told," says Ahn, "about 200 million CAPTCHA squiggles are typed in a day. Each time you type one of those you waste about ten seconds. If you multiply that by 200 million, that means humanity as a whole is wasting around 500,000 hours every day filling out these annoying CAPTCHAs."

So Ahn started wondering if there was a better way to make use of all this time and energy, a way to turn those ten seconds of waste into actual work. "What if," says Ahn, "there was some giant task that humans could do that computers could not that can be broken down into ten-second chunks?" This was the birth of reCAPTCHA, a website that serves a dual purpose, both helping to distinguish bots from humans while simultaneously helping to digitize books.[15]

Normally, we digitize books by scanning pages into a computer; next, an optical character recognition program runs through this text, attempting to turn images into actual words. Sometimes this works great; other times, not so well. The big problem is with old books, especially ones whose pages have yellowed. On average, for books written more than fifty years ago, computers can make out only about 70 percent of the text. That remaining 30 percent—that's where reCAPTCHA comes in. When the computer can't recognize a word, it sends it out as a CAPTCHA—meaning the next time you're typing in drunken letters into your computer, know that you're actually helping digitize the world's libraries. And fast. Ahn's dual-use crowdsourcing platform is digitizing over 100 million words a day—the equivalent of 2.5 million books a year.

And Ahn didn't stop there. "I started wondering about how we could translate the web into every major language. It's a big issue. More than 50 percent of the web is written in English, and less than 50 percent of the world's population speaks English."

But how to translate the whole Web? "You can't do it with fifty or a hundred translators. But what if we could get, say, 100 million people to help translate the Web into every major language. It's a great idea, but how are you going to motivate these people to do this work for free? You can't pay 100 million people. And even if I could, there just aren't that many bilinguals in the world."

That's when Ahn and his colleagues hit on another dual-use idea, realizing they could teach people new languages at the same time as these people were translating the web. "At any one time," says Ahn, "there are about 1.2 billion people out there trying to learn a foreign language. Stuff needs to be translated, so why can't we get the people who are learning a foreign language to translate this stuff for us?"

This was the birth of Duolingo, both a language education website and a translation game that really works. "It's been super-successful," says Ahn. "People on Duolingo learn a foreign language as well as they do with any other language translation program. But because they're translating real content (say, *New York Times* articles), it's inherently interesting—people are motivated by the quality of the content. And because we're using multiple sources for every translation, the results we get are as accurate as those done by professional language translators."

And far, far cheaper.

For example, right now, only 20 percent of Wikipedia exists in Spanish. If you were going to go out and hire translators, the cost of translating the remaining 80 percent would be roughly $50 million. It would also take years and years and years to finish the job. Duolingo, meanwhile, can do it for free, in about five weeks, with about 100,000 users. And as of today, the site has about 300,000 users.

So, while the point of the Freelancer and the Tongal case studies were to explore two different crowdsourcing platforms that offer today's entrepreneur astounding leverage, the point of reCAPTCHA and Duolingo is the inverse—an example of the kind of crowdsourcing platform a bold entrepreneur might be interested in creating, the kind that both makes money and betters the world at the same time.

How to Crowdsource

As you can see from our case studies, crowdsourcing is a diverse and growing field, with more novel applications being dreamed up every day. So, before we dive into lessons learned, to give you a better sense of what's going on, I've broken this section into four of the most common uses for crowdsourcing and provided a short explanation for each.

1. Crowdsourcing Tasks

Tasks are work. Crowdsourcing tasks means getting someone somewhere to do the work for you. In most cases, you pay only if you like the result. In other cases, you can specify what you're willing to pay per task or let the crowd marketplace compete for your business. In this taxonomy, tasks come in two basic flavors: micro and macro.

Microtasks are bite-size, well-defined chunks of work that can either independently solve a small problem or, when combined with many other microtasks (for example, reCAPTCHA), collectively solve much larger problems. This means that one of the most important questions to answer when approaching crowdsourcing is whether the work can be broken down into smaller, simpler units. If so, what is the simplest microtask that can be defined and distributed? For example, a while back I wanted to determine whether *Time* magazine cover articles have gotten more negative during the past sixty years. At first I had my executive assistant start in 1945 and group articles into positive, neutral, or negative categories. After a day's work, she had barely put a dent in the problem. That's when I decided to turn to the crowd. By offering $0.05 per categorization, I got the entire 65 years' worth of issues, roughly 3,000 in total, done for under $200.

I used Amazon's site Mechanical Turk (www.mturk.com) to get those magazine covers analyzed. While MTURK isn't all that useful for more complicated jobs, it is where to go to get simple, quick tasks done fast. Aggregation and classification jobs tend to be popular uses. Aggregate photographs of red trucks, for example, or write product descriptions, or perform sentiment analysis exercises on thousands of Tweets. Requesters (you) post tasks known as HITs (human intelligence tasks) while workers (called providers) browse among existing tasks and complete them for a monetary payment.[16]

Another microtask site that I've previously relied upon (and with great result) is Fiverr (www.fiverr.com), an online marketplace offering microtasks starting at $5. Typical services include voiceovers, animations, crafts, promotional videos, and art. Offerings can get wacky, for example: "I will print and hand out 500 flyers for you in Toronto, Canada, for $5" or "I will draw you in my caricature style for $5." You can scan the site for services of interest to you, or make a request. I've had my hand-drawn sketches turned into great digital art for five dollars.[17]

In contrast to the above microtasks, macrotasks are jobs that (a) can't be broken down, (b) can be done independently, (c) require some type of specific skillset or thought process, (d) are additive and

dependent on already completed work for the task, and (e) take a discrete, fixed amount of time to complete. There are a number of different companies that allow you to crowdsource macrotasks, with the
aforementioned Freelancer.com being the largest. It's also important
to remember what Freelancer.com founder Matt Barrie said about the
diversity of work being done: "This isn't just plumbing and pest control. We've got PhDs on the site. I've seen both quantum physics and
aerospace jobs handled perfectly."[18]

Even better, Freelancer.com is only one of a myriad of macrotask
sites. For a detailed list of the latest sites with examples of how to use
them, please see www.AbundanceHub.com.

2. Crowdsourced Creative/Operational Assets

An asset is anything that provides value to you and your business—
that is to say, applications, websites, videos, software, designs, algorithms, marketing materials, physical goods, machinery, and technical
plans. To understand how to crowdsource assets, I've broken things
into two different categories: *creative* and *operational assets.*

Creative assets include a wide variety of design-based assets such
as logos, videos, website designs, CAD models, marketing plans, and
advertising plans. We've already mentioned two of my favorite creative asset development sites: Tongal (www.tongal.com), which can
make you a TV or Internet ad in weeks instead of months and at a
cost that is usually about a tenth the industry average, and 99Designs
(www.99designs.com), which provides crowdsourced graphic design
(logos, apps, web pages, infographics, blogs, and more). I've used
99designs repeatedly and have found that contests usually yield
between 25 and 400 entries, depending on purse size. Even better, if
you don't like any of them, 99Designs will provide a full refund.

Operational assets, meanwhile, are those things required for business to run effectively. For example, if you are running a software business, these assets include the algorithms powering your software, your

database architecture and server implementation, technical designs, models, and frameworks that organize deal flow and customer acquisition strategies, and so on. A number of companies allow you to crowdsource the creation of operational assets. In fact, doing so is one of the keys to becoming a data-driven, exponential organization.

A great example of this is TopCoder (www.topcoder.com). You've probably heard about hackathons—those mysterious tournaments where coders compete to see who can hack together the best piece of software in a weekend. Well, with TopCoder, now you can have over 600,000 developers, designers, and data scientists hacking away to create solutions just for you. In fields like software and algorithm development, where there are many ways to solve a problem, having multiple submissions lets you compare performance metrics and choose the best one.

Or take Gigwalk, a crowdsourced information-gathering platform that pays a small denomination to incentivize the crowd (i.e., anyone who has the Gigwalk app) to perform a simple task at a particular place and time. "Crowdsourced platforms are being quickly adopted in the retail and consumer products industry," says Marcus Shingles, a principal with Deloitte Consulting.

Retailers and consumer products manufacturers have a challenging time obtaining vital information about how their products are being sold, merchandised, and priced. Is the product in stock? Is the price correct relative to the in-store promotion? Is the display and promotional signage in place? What are competitor products selling for? These are all very key causal data variables that can be converted into insights that a retailer and manufacturer can use to optimize inventory, promotions, and overall sales. It's not always cost effective to have store employees monitor this data, nor is it efficient to send the manufacturer's sales reps into each store to check. Crowdsourced platforms utilize the everyday consumer instead. The consumer gets paid a nominal fee, say five dollars, for taking five minutes to take a photo of a shelf, and the retailer and manufacturer only pay for the five minutes of time needed to collect a specific data point. In our pilots, on average, the crowd was able to deliver this data in less than an hour, and for five to eight dollars per request, across thousands of posted tasks. That data is then brought back into the retailer's and manufacturer's operational systems, in which data visualization techniques are used to make sense of the information collected so timely decisions can be made.[19]

Of course, in this age of exponentials, we are generating more data than ever before. Unfortunately, not everyone knows how to tease out valuable insights from this deluge. Enter companies like Kaggle (www.kaggle.com) and TopCoder (www.topcoder.com), both of which are crowdsourcing, data-mining competition platforms that allow you to define your goal/desired insight, set a monetary prize, upload your data, and watch as hordes of data scientists (tens of thousands, to be exact) figure out the best way to sort through it. The best algorithm wins. The reward levels vary from kudos or zero dollars to hundreds of thousands of dollars from bigger companies.

And for exponential entrepreneurs, not relying on the advantages of data is no longer an option. Deloitte Consulting's Chief Innovation Officer Andrew Vaz explains: "As Big Data overwhelms traditional computing and analytical tools, the combination of A.I. and Big Data will create an insights 'arms race,' where competitive advantage will be dominated by individuals and organizations that capitalize on these emerging technologies."[20]

3. Crowdsourced Testing and Discovery Insights

Insights are invaluable to your business. They can shape the goals and operations of the entire company, dramatically improve and optimize performance, and provide you with counterintuitive ideas or hidden data for a strategic advantage over competitors. When it comes to crowdsourcing insights, there are two main variants: *testing* and *discovery*.

Testing-based insights often come from examining existing assumptions and current best practices. These include surveys, A/B testing, representative sampling, customer feedback, case study abstractions, and focus groups, among others. When running a test, focus on asking one specific question and using the data or resources at your disposal to frame the question appropriately.

For example, if you are in the world of software development, you know that testing your wares can be incredibly tedious, difficult, and time-consuming. There is no room for error, so you have to have as many people as possible hunting for bugs before launch. No problem. uTest (www.utest.com) provides a massive community of "professional testers" who run functional, usability, localization, load, and security tests on your code. By leveraging the crowd and data from previous tests, they optimize and simplify the process, making it less expensive, while offering reduced churn rates, better functionality, and quicker time to market.

Creatives are also getting in on the insight game. Take Reverb-Nation (www.reverbnation.com), a music distribution, publishing, and crowdsourced testing platform. Say you're an aspiring musician. You've produced a few songs, but before spending money on paid advertising or management, you want to see whether anyone actually likes your music. Now you can have songs rated and reviewed long before you actually go to market.

The other side of this insight equation is the crowdsourcing of *discovery-based insights.* This can mean a few things. You can ask the crowd for their interpretation of a particular problem or question. For example, Genius asks the crowd to annotate song lyrics. Kaggle outlines a problem that its community solves with original algorithms. Or perhaps more simply, you can provide a platform for the crowd to come up with their own ideas and inventions, as Quirky does with its invention network and Threadless does with its T-shirt design competitions.

Discovery-based insights can be as simple as asking the crowd for answers and paying attention as the best solutions, designs, and inventions bubble their way to the top. As a personal example, in the spring of 2014, in preparation for rereleasing *Abundance* as a paperback, I asked the crowd to help me discover new evidence of abundance that I could include in the appendix. People emailed me their data, charts, and graphics to evidence@diamandis.com (new submissions are always welcome). The response was extraordinary and yielded considerable evidence of our continued forward progress.

So there you have a top-level overview of the crowdsourcing space. Since this is an area that will change rapidly as new players enter and new AI capabilities come online, I wanted to point you to a few industry websites worth visiting that will keep the taxonomy fresh and up-to-date.

AbundanceHub.com: This is the place I'll be posting my own experiences, providing updates on my lessons learned (successes, experiments, and failures) and working with entrepreneurs interested in creating wealth while creating a world of abundance. Abundance-Hub content is driven by the Abundance360 community, my mastermind group of entrepreneurs whom I've committed to coach over a twenty-five-year period.

Crowdsourcing.org: The industry-leading resource for everything crowdsourcing. Known as one of the most influential and credible authorities in the crowdsourcing space, they are recognized for their in-depth industry analyses, definitive crowdsourcing platform directory, and unbiased thought leadership. Their mission is to serve as a complete resource of information for analysts, researchers, journalists, investors, business owners, crowdsourcing experts, and participants in crowdsourcing platforms.

Crowdsortium: Using the crowd to dissect, organize, and collectively move the young and evolving crowdsourcing industry forward, Crowdsortium helps organizations find, evaluate, and execute new ideas by working with online crowds and providing events, meet-ups, resources, and guides. Crowdsortium was formed by a group of industry practitioners that have the mission of advancing the industry through best practices, education, data collection, and public dialogue.

4. Crowdsourcing Best Practices

While there are way too many crowdsourcing platforms to get into the nitty-gritty of utilizing each, in our research we have identified twelve best practices to apply in almost all cases.

i. DO YOUR RESEARCH

These days, almost anything you want done can be done online. Freelancer.com alone has experts available in six hundred disciplines.[21] The point here is whenever you need something, instead of defaulting to your normal fulfillment process, try leveraging the power of the crowd to do it faster, cheaper, and better. Define your crowd, get familiar with available platforms, and then pick the right one. "If you're going to start a company, it's never been easier and it's never been cheaper to do so," says Matt Barrie. "All the tools you need to build an Internet company today are basically free—all the software: Linux, MySQL, voice-over-Internet protocol, Gmail, and so on. The best thing I would suggest you do is browse the projects on Freelancer. com, and look at the mobile phone section or the web development section or whatever your area might be, and see what other people are doing and how they're wording their projects and what they're paying. Start from there."[22]

ii. JUST GET BUSY

The most consistent advice we received during our research was perhaps the simplest: Just get busy. In most cases, it's free to sign up and post a project. People from all around the world will start bidding on it. And once you start talking to them and looking through their samples of work, they'll give you ideas. They'll tell you, "Hey, I've done similar sorts of projects. Why don't you do it like this or why don't you do it like that?" and so forth. Really, as with all things in entrepreneurship, it's really just a matter of giving it a go and proceeding through trial and error.

iii. TURN TO THE MESSAGE BOARDS

Completing a crowdsourcing project can be really tough your first few times. Each platform is different and there are confusing elements in all of them. For platform-specific guidance, turn to the site's community forums for help. Experts and forum moderators will come out of the woodwork to give you incredible tips and guidance throughout the process. And all that help comes free of charge.

iv. ESTABLISH CONTEXT AND BE SPECIFIC

Don't expect the crowd to understand the core philosophy of your business. What is critical is to give people the context of the project and supplementary resources to consult should they want more background information. A properly established foundation means people spend less time trying to guess your desires and more time delivering exactly what you want.

v. PREPARE YOUR DATA SET

For non-design-specific tasks, there is usually a data set that you must submit for analysis, categorization, and so on. If you don't have a perfectly formatted and ready-to-go .csv file, you can turn to the crowd for help. Most of the time crowdsourcing workers have actually crowdsourced projects themselves, so they know all the best ways to prepare your data for this process.

vi. QUALIFY YOUR WORKERS

Unfortunately, crowdsourcing does have the potential to create undesired results. The quality of the results can sometimes be inadequate, and crowdsourcing is not sheltered from the scammers and bots lurking on the Internet. Luckily, qualifying your workers and curating a trustworthy work force can help you avoid these issues. To qualify a work force, simply put out a few very simple and inexpensive requests to see how quickly and accurately the job gets done. For example, if you have one hundred images you need created, and a dozen crowdsourced workers to choose from, rather than choosing a single graphic artist immediately, consider taking a few of your images and asking a few freelancers to show you their style and speed. Then choose the best to give the entire job to for completion. Just this sort of quick prequalification can save you considerable heartache and time.

vii. DEFINE CLEAR, SIMPLE AND SPECIFIC ROLES

The more clear you are about the role you want your crowd to play in the project, the better the results. If you want creative solutions, tell people. If it's practical solutions, make that clear. Ambiguity doesn't work well in crowdsourcing. Make sure you think through every element of the product or service you need, and be ready to field questions, concerns, or confused comments. These are almost inevitable.

viii. COMMUNICATE CLEARLY, IN DETAIL, AND OFTEN

"Remember," says Freelancer's Matt Barrie, "you're often working with someone on the other side of the world. If you just write a one-line sentence—'I need a website'—that could mean anything. The better the description you provide of what you want and the less room [there is] for interpretation, the better outcome you'll get."[23] Many platforms allow you to communicate with the crowd during the campaign—

choose one of these and communicate often. This collaborative strategy is critical for producing the best results.

ix. DON'T MICROMANAGE; HAVE AN OPEN MIND
FOR NEW WAYS OF THINKING

This may seem counterintuitive considering the previous two bits of advice, but often the best results from crowdsourcing come from unexpected angles. Sure, if you're dealing with a microtask, then creativity isn't welcome, but for larger projects—like, say, a Tongal video—give people the space they need to surprise you and they will. "Don't do what's expected," DeJulio said. "Try something new. Allow the crowd to come up with wild and crazy bold ideas that might complement the genius of your brand. Trust the process; don't try to re-create what you're doing normally. The goal is to get something new and fresh."[24]

x. PAY TO PLAY—THAT IS, GO FOR QUALITY FIRST, THEN PRICE

Crowdsourcing is cheap, yes, but you shouldn't be. If you're running a 99Designs contest, for example, the difference between the number of design submissions you'll get for $199 and $299 is far greater than the $100 you spent. "Put a budget range down," says Barrie. "Then it's a free market. The freelancers will bid on the project and tell you what they want to be paid. It may be an hourly rate, if it's that sort of a model, or it may be a fixed price. You can look through the bids, but the most important thing is go for quality first, because the price is going to be so cheap anyway that you're going to have tremendous cost savings. That gives you tremendous leverage in terms of what you can do with your starting capital."[25]

xi. PREPARE FOR THE FLOOD

Crowdsourcing offers a counterproblem to the classic dilemma of "not enough good ideas." The issue with crowdsourcing is the opposite.

You're going to be flooded with good ideas. Prepare for the deluge. You'll need to be clear about your big objectives, but remain open to new ideas—that is, new paths to your big objectives. This deluge is the true advantage of crowdsourcing.

xii. BE OPEN TO NEW WORKING METHODOLOGIES

"The other day I was in London," said Barrie, "and I met a financial analyst working from home doing financial models for pension funds on things like infrastructure projects. He needed a mathematician to develop these models in MATLAB to be able to do his research and present his findings, so he hired a PhD student in Pakistan to do the work. They set up a chat on Skype. The streaming video quality to somewhere like Pakistan is now unbelievable. It was just like the guy was in the room with him. He'd get up in the morning, have his cup of tea, sit down, put the iPad there, do the video call, and then they'd sit there and talk all day as if they were in the same room together. The ability to communicate with anyone on the planet is getting better and better. That means the ability for us to work with anyone on the planet is fantastic."[26]

So there you have it. A quick overview on the exponentially exploding world of crowdsourcing, today's poor man's version of artificial intelligence. More incredibly, today the exponential world is starting to overtake the crowd. Recently, an AI company called Vicarious, which is backed by such investors as Elon Musk, Jeff Bezos, Mark Zuckerberg, and Peter Thiel, announced that their machine learning software achieved success rates up to 90 percent on CAPTCHAs from Google, Yahoo, PayPal, Captcha.com, and others.[27] So stay tuned, since even the crowd can eventually be dematerialized and demonetized.

But one use of the crowd that AI is unlikely to disrupt in the near term is the ability of people from around the world to send you cash to underwrite your ideas. For more on the billions of dollars flowing into the crowdfunding space and the best way for you to capture those dollars, let's turn to our next chapter.

CHAPTER EIGHT

Crowdfunding

No Bucks, No Buck Rogers

The Money Question

When I started working on this book, my research team and colleagues at Singularity University polled a thousand would-be entrepreneurs and a thousand established entrepreneurs about the greatest barriers encountered when starting a business. Raising money topped almost everyone's list. No surprise, right?

While the numbers move up or down a little each year, at any given time, there are roughly 27 million US businesses in need of capital.[1] According to the US Small Business Association, lack of capital is the main reason why 50 percent of new businesses fail within their first five years of operation. Yet, despite this clear need, 23 percent of the companies are so daunted by the prospect of raising money that they don't even try, while another 51 percent get turned down for their effort.[2]

All of this, though, is starting to change.

Historically, our access to capital has been limited by our access to people. While reaching out to friends and family is how most entrepreneurs get started, by definition, this is a very limited group with

potentially limited means. Next on the list are traditional investors—
folks like angels, super-angels, and venture capitalists. But, in my experience, many of these professional investors are too narrowly focused
and shortsighted for bold adventures. Yet in today's hyperconnected
world, entrepreneurs have instant access to millions of potential backers and a billion-plus potential customers.

Crowdfunding is the exponential crowd tool that lets you tap this
new resource, allowing you to mine the world for like-minded individuals and fast-track passion projects like never before. The first crowdfunding platforms hit the scene in the latter half of the last decade.
These early iterations were tools primarily used by filmmakers and
musicians looking for ways to fund projects without the backing of a
major label/studio, but it didn't take long for a far wider range of entrepreneurs to get involved.

And stay involved.

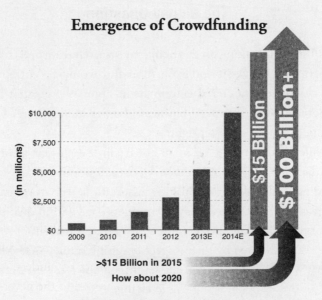

The Growth of Crowdfunding

Source: www.forbes.com, www.entrepreneur.com, www.gsvtomorrow.com

"E" refers to "Estimated", as in estimated size of the market.

In less than seven years, crowdfunding has become a significant economic engine. There are more than 700 crowdfunding sites online today, funding every variety of project under the sun.[3] And that number is expected to double over the next few years. Globally, the total funds raised have followed an exponential curve from $530 million in 2009 to $1.5 billion in 2011 to $2.7 billion in 2012.[4] By 2015, experts predict a $15 billion crowdfunding market, which with the passing of the JOBS Act and the addition of equity crowdfunding to the scene could become an incredible $300 billion marketplace over the coming years.[5]

What has also increased is our confidence in the process. In the early days of crowdfunding, when a film was prefunded by the crowd, this was a clear sign that people actually wanted to see the film. The same holds true for a product or service. Consider +Pool, which raised over $270,000 on Kickstarter.[6] Its service? A filtered swimming pool floating in the middle of the East River. Who knew New Yorkers wanted to swim that badly? The fact is, the people behind +Pool suspected it, and their Kickstarter campaign confirmed it. That's the unique power of crowdfunding.

For the entrepreneur, this kind of social proof is invaluable. It's also lucrative. Crowdfunding expert Candace Klein believes that no matter who you are, there's usually an untapped $100,000 floating around your social network. "I think this is true no matter what social strata you come from. I'm a perfect example. I was raised in a trailer park. The first time I tried to raise money from my network—it was almost a joke. No one I knew had any money. A lot of my friends couldn't even pay their bills on time. It took two years, but I raised $200,000. And that was without the help of crowdfunding. Today these platforms allow you to speed up this process and expand your reach and allow you to market to a completely different audience of funders."[7]

But crowdfunding does more than put an end to the never-ending fund-raiser's dilemma, it also provides entrepreneurs with a significant psychological boost: the ability to start by starting.

It's all about momentum. With any project, the most dangerous period is the zest-sucking stretch between "I've got a neat idea" and "I'm actually doing real work on that idea." In my case, it took nearly a decade to raise the millions needed to fund the first XPRIZE the old-fashioned way. But it took just thirty-four days for me and the Planetary Resources team to crowdfund the $1.5 million to underwrite the launch of the ARKYD space telescope (a story we'll come back to in a little while).[8] In other words, crowdfunding is the antidote to this energy-draining stall, allowing exponential entrepreneurs to immediately get into the game.

Here's how to play.

The Types of Crowdfunding

There are four main types of crowdfunding, each based on what the investor receives in return for helping to fund a campaign: donation, debt, equity, and reward.

1. *Donation.* This is simply the digital version of traditional charity. Donors get little beyond gratitude and a receipt to claim on their taxes. Examples include DonorsChoose, GlobalGiving, and Causes.

2. *Debt.* Sometimes referred to as microlending or peer-to-peer (P2P) lending, this variety of crowdfunding involves an entrepreneur asking the crowd for a loan, and, in return, repaying that loan with interest. Examples include Kiva and LendingClub.

3. *Equity.* This is the newest type of crowdfunding, a development made possible by recent changes to the US Securities and Exchange Commission (SEC) regulations.[9] In equity crowdfunding, entrepreneurs can now sell equity in their company online, asking investors for cash in return for stock. Examples include Crowdfunder, Startup Crowdfunding, and (for those who have already raised their first $100,000 in capital) AngelList.

4. **Reward or Incentive.** The funder sends money to support the creation of a product or service that inspires him or her and in return receives a reward. Simple as that. Send $25 and get a T-shirt. Send $100 and get a copy of the product you're helping to fund (technically, a presale). The numbers vary a bit, but in general, reward-based crowdfunding is 60 percent more effective than straight-ahead donor funding. Examples include Indiegogo, Kickstarter, and RocketHub.

So which type of crowdfunding is right for your project? Well, of the four, donation funding is fine for social causes and political campaigns, but is not often utilized for entrepreneurial ventures. Debt funding, meanwhile, is best for local projects that benefit a community, such as helping someone open a new restaurant, hair salon, or retail shop, but has not performed well for larger entrepreneurial ventures. If you're a business looking to expand locally, this platform is for you. Otherwise, look elsewhere.

Equity funding is the most recent category of crowdfunding, becoming possible only because of the 2012 Jumpstart Our Business Startups (JOBS) Act, which allows new businesses to use crowdfunding to raise early-stage equity-based financing. "For the first time in nearly eighty years," says Chance Barnett, CEO of the equity site Crowdfunder, "private start-ups and small businesses can raise investment funding publicly, using sites like Facebook or Twitter to help spread the word and taking in investment online via equity crowdfunding sites who power the investment process in a more open and collaborative way."[10]

The potential here is tantalizing, with estimates running as high as $300 billion for the future size of the equity market. But even today, with equity just getting started, the amount of money involved is already considerable. As of July 2014, Crowdfunder has processed $105.2 million in deals with more than 11,000 companies listed and 62,000 investors registered on the site.[11]

AngelList is another equity platform that's getting a lot of attention—and for good reason.[12] Started in 2010 by Babak Nivi and Naval Ravikant, AngelList is a platform for startups to meet angel investors, and vice versa. Investors and startups can create profiles, list their investments, and connect to one another. Participants are some of the best in the business. For example, Uber, the ridesharing service discussed earlier, not only raised their first $1.3 million on the site, but also met investor Shervin Pishevar, who later lead a $32 million Series B for Uber at Menlo Ventures (Pishevar has also become one of the biggest investors on AngelList). The best news: you don't have to invest millions of dollars to get into these deals. In 2012, AngelList partnered with SecondMarket to give smaller accredited investors the chance to invest as little as $1,000 in start-ups alongside top technology investors.[13]

Yet, despite equity's enormous potential, the focus in this chapter is going to primarily be on the fourth category: reward-based crowdfunding. We have chosen to place our attention on reward campaigns because equity crowdfunding is still too new and lacks the hard data required for accurate strategy suggestions. Debt crowdfunding, meanwhile, remains primarily a local mechanism and unsuitable for the bold. But the real reason for this focus is that reward-based crowdfunding already has a long entrepreneurial track record of success, proving itself effective for funding creative projects (movies, CDs, books) and actual products (watches, telescopes, even bioengineered plants). More importantly, as we'll see in the following examples, it is a tool that continues to expand its reach.

To these ends, we're going to examine three different reward-based funding efforts. The first is Pebble Watch, which is both one of the most successful crowdfunding campaigns ever and a really great example of how a small team of entrepreneurs can fund and launch a new product.[14] The second case study is Let's Build a Goddamn Tesla Museum, which highlights the expanding abilities of individuals to fund far more unusual projects, showing how the combination of enormous passion and the right partners can make all the difference. The final example is the ARKYD Space Telescope, a campaign run by my company Planetary Resources, which helped us start and forge an enormously passionate community of space enthusiasts—generating the kind of support that is absolutely required by this kind of future-forward project.[15]

One quick clarification: these examples have been kept intentionally short because they'll again be followed by a lengthy how-to section—the real meat of this chapter. It's here we'll break down everything you need to know to get started, providing information drawn from four sources: a meta-analysis of all the major crowdfunding guides that have appeared in the past few years (twenty-six in total); lengthy interviews with the founders and CEOs of major crowdfunding companies such as Indiegogo, RocketHub, and Crowdfunder; lengthy interviews with entrepreneurs who have run incredibly successful campaigns (for example, Eric Migicovsky, creator of the Pebble Watch campaign); and finally, my own personal experience raising $1.5 million via crowdfunding, which at the time was the twenty-fifth most successful Kickstarter campaign ever. In total, this chapter will enable you to design and launch a reward-based crowdfunding campaign, something that every exponential entrepreneur should plan on experimenting with and many will find core to their mission.

Case Study 1: The Pebble Watch

How do you ride a bicycle and answer your phone at the same time? This was the question Eric Migicovsky was trying to answer in 2008. Migicovsky was an engineering undergrad enrolled at the University of Waterloo in Ontario, Canada, then spending a year abroad studying industrial design at Delft University of Technology in the Netherlands. He was also riding his bicycle everywhere. Everyone rode in the Netherlands, and Migicovsky was a fast convert to these two-wheeled ways.

But he was also a frustrated convert. Whenever Migicovsky was on his bike and his cell would ring or a text or an email would arrive, he faced that dread decision: Stop to answer or miss the message. Some of those messages were important. What would really help was a way to know at a glance who was on the other end of the line (or who was sending that email or text) and if it was worth pulling over to take the call or write a response. There were a few devices around that could do that—the Fossil Wrist Net, for example—but all came with hefty price tags. Migicovsky wanted something affordable—call it a smarter watch for the common man. That's when he decided to build his own.

Back in Canada, during his final year of college, Migicovsky cobbled together his savings, winnings from a business plan pitch competition, and a $15,000 loan from his parents, then assembled a small team and built a prototype. The inPulse was born. It's a smartwatch. It tells time. It syncs up with mobile devices. But mostly it's an at-a-glance alert system—using its big flat screen to notify you about incoming cellular messages.

The inPulse developed a core fan base, but because the first iteration worked only with a BlackBerry (this was 2008 and Migicovsky, like BlackBerry, was Canadian), it didn't go big. Yet there was enough early initial traction that Migicovsky decided to move the project to Y Combinator in Silicon Valley, which is also where he found the seed money to start manufacturing an updated version of the inPulse. And that's when he hit the wall.

Some great customer feedback had led to further rounds of design improvements, which resulted in an entirely new watch, the Pebble. It's a great watch. It syncs up with iPhone and Android, runs apps, and allows users to check their calendar. "Basically," as Migicovsky later told *Inc.*, "the smartest watch ever."[16]

Unfortunately, finishing this smartest watch ever required an additional $200,000 in funding, but Migicovsky and his team had problems raising the money. They met with VC after VC—many of whom had funded them before—but couldn't ink a deal. No one wanted to take the risk. With just enough cash to keep going for another two months, Migicovsky and partners turned to Kickstarter for one last-ditch effort.

Migicovsky started making phone calls. He rang up the team at

Supermechanical in Austin, Texas, who had "accidentally"—meaning their original goal was just $35,000—raised over $500,000 for Twine, a Wi-Fi sensor that connects everyday objects to the Internet. Migicovsky grilled them on the basics: how to make a video, how to post, what to expect. Then his team embarked on additional research, studying hundreds of winning campaigns and the strategies behind them. In the end, the Pebble crew settled on a pitch video that was fairly low-tech, sort of seventies funk meets geek dirty talk—"we created a prototype from cell phone parts," etc.—while their reward levels were designed around customer feedback. Tip to tail, the whole campaign creation process took only six weeks. Then again, it had to, as they were out of money.

How a crowdfunding campaign performs in those first few hours after launch matters considerably, so the Pebble crew brought in the tech blog Engadget as their exclusive launch media partner. At 7:00 AM, the campaign opened for business and an Engadget article hit the web. The Pebble crew held their breath.

"We set a fundraising goal of $100,000," said Migicovsky, "but that wasn't actually the truth. We really needed to raise $200,000 to build and deliver the watches. I made a pact with the team—we set $100,000 as our target, but decided that if we didn't actually raise $200,000, we would return everyone's money because we just wouldn't feasibly be able to make it."

What happened next surprised everyone. "Two hours after the Engadget article went live, we hit $100,000," recalls Migicovsky. "We hit $200,000 about two hours after that. In the first twenty-eight hours we raised a million dollars. The first day was spent mainly in awe at what we had started. By the end of the campaign, on May 18, 2012, we had passed $10 million. Our backers came from all over the world. This was a result we never expected."

Also unexpected were their final totals. By the numbers, Pebble raised exactly $10,266,845 from 68,929 backers in thirty-seven days—a world record at the time. And the results since have been nothing more than stellar. After being turned down just a year earlier for $200,000 in funding, Migicovsky successfully raised an additional $15 million in VC backing on the heels of their successful campaign. Better yet, Pebble sold more than 400,000 watches in their first twelve months, beating iPod's first year (they shipped 394,000).

Case Study 2: Let's Build a Goddamn Tesla Museum

Popular web comic artist Matthew Inman, aka "the Oatmeal," has been a long-time fan of inventor Nikola Tesla. And why not? This was Tesla, the man who invented alternating electric current (though the credit went to Edison), wireless radio (though the Nobel Prize went to Marconi), and, well, X-rays, radar, hydroelectric power, and large portions of the transistor (again, in each case, no credit). Tesla even experimented with cryogenics—fifty years before the field actually had a name. Thus, when Inman found out that Wardenclyffe, Tesla's final laboratory (located in Shoreham, New York, and where the inventor attempted to create a power station that would provide the world with *free* electricity), was up for sale—with an offer already on the table to buy the land, tear down the lab, and put up retail stores—he had to do something.

Partnering with the nonprofit Tesla Science Center (which had been trying to buy this land for eighteen years), Inman turned to the crowd-funding platform Indiegogo. Next he created a long, funny comic about Nikola Tesla—who he was, why he was important, and why the world needed to buy that property and build a museum to honor his legacy. Humorously titling the campaign Operation: Let's Build a Goddamn Tesla Museum, Inman released the project in August 2012.[17]

Within a day, it went viral. Over the course of its first week, it raised $145,000 per day, $6,000 per hour, and $100 per minute. Then things got really interesting. At one point during the campaign, contributors were donating over $1,000 a minute. By the end of the month, they had shattered their goal of $850,000, raising over $1.3 million from 33,000 backers in 102 countries.

Like the Pebble Watch campaign, the Tesla Museum campaign marked a turning point in crowdfunding. What made it unique was the fact that it was a nonprofit initiative (generally, nonprofits don't do as well on crowdfunding platforms) without any real product to offer (meaning, unlike the Pebble watch, backers would never get a gizmo in the mail). As a result, the Tesla Museum pushed boundaries, becoming the most successful campaign on Indiegogo and opening the door for the crowdfunding of larger, more conceptual projects. The Oatmeal, of course, commemorated the whole thing with a new ending to his comic, writing: "Mr. Tesla . . . We're sorry humanity forgot about you for a little while. We still love you lots. Here's a Goddam Museum."

Case Study 3: The ARKYD Space Telescope—Access for Everyone

When Eric Anderson, Chris Lewicki, and I launched our asteroid mining company Planetary Resources, we knew we needed a powerful community behind us. When you're doing something as radical as asteroid mining, having a group of passionate supporters is downright necessary. The question was how we could actively and authentically engage the public in our mission to explore space. After all, space is expensive, difficult to access, and—as we've learned by now—stubbornly impenetrable. Furthermore, people care most about things they can have an immediate impact on, but from the beginning of humanity's extraplanetary exploits, most everything we've done in space has been done by a small group of experts and has taken decades to get off the ground. We wanted to stimulate people's interest now, hopefully by providing them with a way to participate firsthand. This was when we hit upon the idea of crowdfunding the first ever crowd-controlled space telescope, the ARKYD.

Thus began a four-month journey as we started researching, planning, and developing our Kickstarter campaign. We appointed one member of our team, Frank Mycroft, to be in charge of the effort, assembled a group of affiliate launch partners, including notable space and science celebrities such as Bill Nye, Hank Green, and Brent Spiner, and finally hit on the idea of pulling together an army of super activists we dubbed Planetary Vanguards to help us implement and promote the campaign. Our media team got busy shooting, editing, and testing our pitch video. Our technology team finalized the prototype designs, renderings, and prototypes of the ARKYD space telescope so people could see what the finished product would actually look like.

But we had another problem. Most successful "product" campaigns actually had a product to offer. We didn't. We weren't selling a cool new watch; ours was a telescope designed to hunt for asteroids. While we could offer our backers an image taken from the telescope of, say, an asteroid, a galaxy, or the Moon, when we polled our existing community for feedback, we discovered this idea wasn't likely to go viral. Then, in the month right before the campaign launched, one of our team members suggested offering a "space selfie"—a chance for anyone to send a photo of themselves up to our spacecraft, where that image would be displayed on a screen, photographed with the Earth in the background, then sent back down to them (more on this later). Priced at a $25 reward level, we thought this was the perfect solution, and when we tested it, our community agreed.

On May 29, 2013, we held a press conference and launched our crowdfunding campaign. Our goal was to raise $1 million—enough money to launch the telescope into orbit (the actual cost of the telescope was being covered by Planetary Resources). Within the first two days, we'd raised close to $500,000. Our Vanguards proved invaluable, spreading the word and pushing the movement forward. Thirty-two days later we finished with $1,505,366 in funding from 17,614 backers.[18] While this was the largest space-related crowdfunding campaign to date, more important was the community we built. As a company working to do something new and bold, having a group of passionate supporters will prove priceless when we seek to crowdsource future support and solutions.

The Money Solution:
A How-To Guide to Crowdfunding

In the section below, you will find some of the most valuable lessons I've learned while researching successful crowdfunding campaigns and implementing my own. I'll give an overview of the topics, then provide far more detailed analysis in the pages to come. In sharing this knowledge, I hope to enable many more innovators and disruptive entrepreneurs to launch their own wildly successful crowdfunding ventures.

Who Should Do a Crowdfunding Campaign?

While crowdfunding can be an immensely valuable tool to raise capital and grow a community, it isn't for everyone. My research shows that the best crowdfunding campaigns have five key characteristics in common:
- The product is usually in late prototype phases, sufficient to show prospective backers what they're supporting.
- The team is correctly assembled and capable of executing.
- The product is community focused and consumer facing.
- The team has access to a large community of followers who can be pitched directly, or has the ability to marshal significant public relations/media resources to attract attention.
- The product aims to solve a problem, improve an existing product, and/or tell a new story.

Seven Reasons to Consider Crowdfunding

Crowdfunding has a variety of benefits beyond raising capital. I've listed some of the most important ones below. Remember, a great deal of how you design your campaign depends on which of these benefits you desire most.

- *Market validation and real demand measurement.* Perhaps the most valuable benefit is your ability to get real customer feedback on your product—what features they desire, what colors, what accessories, etc. Equally critical, you also find out what they don't want. And unlike surveys or focus groups, here customers are voting with their wallets. Of course, as a bonus, you will also get far more general data—geographic info, price sensitivity, and so forth—that can help you shape strategy.

- *The raising of significant investment capital.* Interestingly, some venture capitalists are requesting that a company run a crowdfunding campaign to validate market interest before making an investment. In this case, a successful crowdfunding campaign has two advantages: providing non-dilutive capital to grow the company in the early days, and allowing the company to command a higher valuation in its venture round. Just look at the results achieved by the Pebble Watch team, raising $15 million in venture funding twelve months after their campaign. Even more impressive, Oculus Rift went from raising $2.4 million on Kickstarter to being acquired by Facebook for $2 billion in just eighteen months.[19]

- *The development of a paying community of customers.* There is enormous value in building and having access to a community of paying customers, yet this is very difficult to do in a normal marketplace, and almost impossible to accomplish before your product is released.

- *Cheap cost-per-customer acquisition.* Acquiring customers through other means is often orders of magnitude more expensive. Moreover, if you execute correctly, not only do you get free advertising, but your customers both pay you to be involved and promote your product to their friends and family.

- *You're passionate about your product.* If you love your idea and want to get it out there quickly, there is perhaps no better way to make this happen fast and potentially at a profit.
- *Public relations benefit.* Success begets success. The positive brand image and media attention associated with crowdfunding success has intrinsic value, putting the company on the map and making future product offerings easier and more lucrative.
- *Cash-flow positive.* Oh yes, there's also the benefit of near-term dollars. Balancing costs and earnings correctly puts cash in the bank for product development.

Execution—Twelve Key Steps

So if one or more of those seven reasons resonates with you, and you're ready to create and launch your own crowdfunding campaign, the following is an outline of the twelve key steps needed for execution.

1. Choosing your crowdfunding idea (product, project, or service)
2. How much? Setting your fund-raising target
3. How long? Setting your campaign length and creating a schedule
4. Setting your rewards/incentives and stretch goals
5. Building the perfect team
6. Sharpen your ax: planning, materials, and resources
7. Telling a meaningful story (and using the right words)
8. Creating a viral video: three use cases, shareability, and humanization
9. Building your audience—the three As
10. Super-credible launch, early donor engagement, and media outreach
11. Week-by-week execution plan: engage, engage, engage
12. Make data-driven decisions and final tips

1. CHOOSING YOUR CROWDFUNDING IDEA (PRODUCT, PROJECT, OR SERVICE)

The most basic question is what should you choose to crowdfund? The short answer is found at the intersection of two primary drivers: First, find something you feel deeply passionate about creating. Second, choose something the crowd is passionate about seeing come into existence. A quick review of sites like Indiegogo, Kickstarter, and RocketHub shows an extraordinary diversity of things getting funded these days (see below). The point is that just about anything you're passionate about already has a crowdfunding history:

Arts/Entertainment

1. Creating a film
2. Writing and publishing a book
3. Launching a new play
4. Opening an art gallery
5. Creating a CD or music video
6. Producing a concert or festival
7. Creating a video game

Charities

8. One-shot charitable projects (a disaster relief project)
9. Starting or growing NGOs
10. Sponsoring a youth sports team
11. Supporting a school
12. Animal initiatives
13. Supporting deserving individuals

For Start-ups and Existing Business

14. Market-testing a prototype
15. A new piece of hardware
16. A new software capability
17. Launching a new service
18. Clothing and fashion companies
19. Starting a digital magazine
20. New food/snack/drink products

If you're considering a number of crowdfunding possibilities but are unsure which one is best, ask your community. Post your idea online (Google+, Facebook, etc.) and get feedback. You also want to pick an idea that is far enough along that people believe you'll be able to pull it off. If your product or project is underdeveloped, people will doubt your ability to make it real, and you're unlikely to get the support you need. On the other hand, if the product is too advanced— nearly finished and ready to ship—why would people want to back it?

2. HOW MUCH? SETTING YOUR FUND-RAISING TARGET

So now that you've settled on what you want to crowdfund, the next question is, how much money do you want to raise? Here again there is a bit of psychological strategy you need to consider.

Crowdfunding is all about incentives. The success of the campaign is wholly dependent on creating early excitement and offering urgent, exclusive, and value-added incentives.

Thresholds. On many crowdfunding platforms, they allow you to run only fixed-funding campaigns—meaning you get to collect the cash only if you reach your stated fund-raising goal. Thus, the most important threshold is the amount needed to reach that goal, but this number can be tricky to estimate. For example, in 2012, Indiegogo found that, on average, campaigns that set their goal between $50,000 and $75,000 raised more money than campaigns that set their goal at $100,000.[20]

In a fixed-funding campaign, setting the goal too high means that even if you raise millions of dollars and excite thousands of backers, you don't get to keep a penny. For example, Ubuntu tried to raise $32 million for a new phone on Indiegogo. While they eventually reached over $12.8 million in funding, they drastically undershot their threshold and all $12 million was returned to contributors.

So how do you set the right threshold? The first thing to remember is that crowdfunding is not where you make a profit on your project. It's where you offset some of your expenses. Notice I said *some* expenses. In most cases, you won't crowdfund all the development costs, but you will recover a significant portion of money that might otherwise have come from investment capital or out of your own pocket.

The second important point is that everyone loves a winner. If people believe your crowdfunding campaign is likely to succeed, it will. In other words, if you appear credible and able to meet your target goal, then people are more likely to back you. Conversely, if people don't believe you, they'll never whip out their credit card. In fact, the research shows that campaigns that reach 30 percent of their goal have 90 percent chance of success (that is, raising the desired amount). In the case of Pebble Watch, Eric Migicovsky needed $200,000 to move forward, but he set a goal of $100,000. In the case of the ARKYD Space Telescope campaign, the cost of building and launching the space telescope was roughly $3 million, but Planetary Resources set a goal of $1 million, with the expectation that these funds would at least help offset launch costs.

The objective here is to figure out the absolute minimum amount that, if funded, would represent enough for you to move forward. When coming up with this number, it is important to remember that there are additional costs built into any crowdfunding campaign: (1) transaction fees, which include both credit card costs at 4 to 5 percent, and platform fees, which also tend to range from 4 to 5 percent; (2) the cost of fulfilling on your rewards. We'll talk more about these later.

Table 8-1. Calculating Your Campaign-Funding Target

What is the minimum you need to raise
to move forward?   _____
+ a 10 percent margin to cover all platform
and credit card fees   _____
+ the cost of all the rewards you need
to fulfill   _____

Total: Campaign-Funding Target   _____

Stretch Goals. So once you meet your original funding goal, what motivates the crowd to keep going? That's where a concept called stretch goals comes into play. These are objectives added to the campaign as you approach your next funding goal. In the case of ARKYD, after we had hit the $1 million campaign target, we set a stretch goal for $1.3 million, then $1.4 million, and ultimately $1.5 million. Basically, each stretch goal promised the users more ways to become involved.

3. HOW LONG? SETTING YOUR CAMPAIGN LENGTH AND CREATING A SCHEDULE

Typical campaigns run from 30 to 120 days, though data from Indiegogo indicates that shorter campaigns (an average length of 33 to 40 days) perform better than longer campaigns.[21] For comparison, the Pebble Watch was a 37-day campaign, while ARKYD was 32 days and the Tesla Museum was 45 days. As you can see from the graph on page 188 (which maps funds against time), with ARKYD there was a huge spike in the beginning, a small spike in the middle (when the campaign exceeded its $1 million target), and a big spike at the end. Such drop-offs in the middle of the campaign are typical and are the reason that stretching the campaign out longer doesn't help.

Pledges per Day: ARKYD—A Space Telescope for Everyone

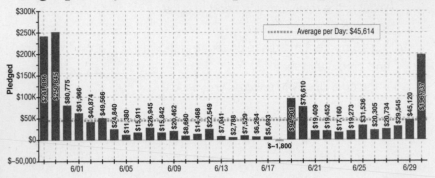

ARKYD Campaign—Pledges per Day

Source: www.planetaryresources.com

When to launch your campaign is up to you, but you should try to give the media a reason to get excited. Tying the launch to a major announcement or anniversary that turns it into a news story can help a lot.

How long will it take you to prepare? Smaller campaigns ($1,000 to $50,000) can probably be done in a month, but if you are looking to raise hundreds of thousands or millions of dollars, much longer-term preparation is critical. Following is my quick overview of how to estimate this for yourself:

Table 8-2. Prep-Time Calculator

Baseline	+30 days
Do you have a team: No?	+30 days
Do you have a community? No?	+30 days
Is your goal less than $50,000? Yes?	+30 days
Is your goal more than $250,000? Yes?	+30 days
Is your goal more than $1 million? Yes?	+30 days
Total Prep Time	___ days

The ARKYD Space Telescope campaign, for example, took us roughly four months (120 days) to prepare. Note: If you're shooting for a significant fund-raising goal, don't be surprised if you delay the launch date once or twice along the way. My advice is: Launch when you are ready. Don't create a false deadline for yourself.

4. SETTING YOUR REWARDS/INCENTIVES AND STRETCH GOALS

Rewards are what backers get for contributing. In our research, lower value perks attract more total contributions, while higher value perks raise a higher percentage of total funds. Though each campaign is different, most sites recommend offering strong perks priced at $25, $50, $100, $500, and $1,000. They report that the $25 perk is the single most claimed reward, representing nearly 25 percent of all perks selected.[22]

Total Revenue by Reward Level

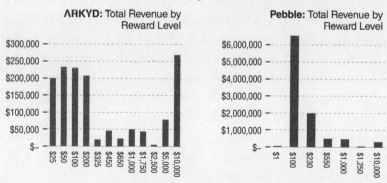

Source: www.planetaryresources.com

The best perks offer rewards that customers would not be able to purchase under any other circumstances, meaning they're unique, exclusive, and authentic to your campaign. Planetary Resources's "space selfie" was something you couldn't purchase elsewhere, had never before been attempted (hence unique), and best of all, was digital—which meant fulfillment costs were essentially zero. We also packaged our offerings, combining space selfies with everything from memberships in the Planetary Society to scholarships for entire schools. At our lowest level, $10, we wanted to encourage our fans to get involved, opt into our email list, and contribute to the community. The key was to make the rewards simple, meaningful, and valuable to our backers.

Reward levels can range from $1 (minimum) to $10,000 (maximum) on Kickstarter (Indiegogo allows higher level perks). It is important to have a low-priced, all-digital, no-brainer reward at the low end because it brings people into your community at no cost to you. This has two further benefits. First, 62 percent of successful campaigns have repeat funders. Get people to sign up. Once they are in the door, you can upsell them later. Second, if your goal is to create community—people whom you can work with and sell products to long after your campaign is over—then you want them involved at any price.

If your goal is to swing for the fences and raise $250,000 or more, it's also important for you to have a compelling $10,000 reward level. We'll discuss this later, but know that $10,000 rewards are a way to get a strong jump out of the gate and help create a super-credible launch.

The final piece of advice should be obvious, but is worth stressing. Ask your community. It doesn't matter what you want; it matters what they desire. So ask them what rewards you should give away. Ask them via email, on Google+, Facebook, whatever. In the early days of the ARKYD campaign, we set up a page where we engaged our existing community, proposed rewards, and asked them to comment and vote for those rewards they liked best.

Scarcity Helps. Take a look at various crowdfunding campaigns and you'll see that most rewards are limited in number. For example, there may be only a thousand $100 rewards or twenty $10,000 rewards. These limited numbers create the illusion of scarcity and a sense of urgency among backers; they'll want to pledge now and not later. The truth, of course, is that you can add additional rewards at any time during the campaign, so if one of your reward levels fills up, you can launch a similar reward at an identical price.

Adding Rewards and Stretch Goals. As I just mentioned, you can add new reward levels to your campaign along the way. This is done for a number of reasons. First, if a reward level sells out, you can add it back in at a similar price point. Second, you may learn that what you thought the public wanted isn't quite what you're offering. Thus, utilize rapid experimentation and make changes when needed. Third, you've met your goal and a stretch goal is needed. This stretch goal warrants a new set of rewards. Sixty-two percent of successful campaigns have repeat funders, and 20 percent of repeat contributions are for perks that were added after the campaign went live.

5. BUILDING THE PERFECT TEAM

Pretty much anyone in Silicon Valley these days will tell you that building a great team is the most important step in building a great product or company. The same applies to building a great crowdfunding campaign, but the components of the team are somewhat unique. Pebble Watch founder Eric Migicovsky built the first iterations of his smartwatch with three friends, all engineers. Yet as soon as Pebble's Kickstarter page went viral, he had to hire an entire external PR team to manage the deluge of attention. And it's not just Pebble Watch. Across the board, and even if you're looking to raise far smaller sums of money, these campaigns are very labor intensive. They require time and effort to plan and execute, and a great team is going to make all the difference.

Our research shows seven key team roles—five are mandatory, two are optional—that must be filled in order to give yourself the highest chance of success. That said, for a campaign in the range of $5,000 to $25,000, it is possible for an individual or a couple of people to pull this off, but if you're looking for six or seven figures, you'll need a much larger team. Here are my recommendations:

The Celebrity (the Face). This person will act as the face of the campaign. He or she will be featured in the main pitch video, will be the voice of your campaign updates, and will lead all other public-facing efforts to garner support. This person has to be emotionally invested in the project, intelligent, eloquent, humble, and genuine. Being funny is an added bonus. The celebrity should also be an expert in the product or service you are offering. He or she will work tirelessly to build momentum in the planning, launch, implementation, and wrap-up phases of this campaign. Translated into standard organizational language, the "celebrity" could be the CEO of your company or possibly one of the more charismatic and passionate members of the founding team. Alternatively, it can be worth bringing in an outsider to lead the public effort—which is sort of what happened with the Tesla museum. The Tesla Science Center had spent eighteen years trying to raise the money to preserve that laboratory, but it was only once Matthew Inman got involved that things got moving.

Campaign Manager and Strategist. The campaign manager fills perhaps the most important role in the campaign. From the earliest days of the planning period, your campaign manager will be running the show. After having done the lion's share of the market research, he or she will lead the development of everything from pledge levels and rewards to distribution channels and partnerships. The campaign manager will plan, organize, and manage the day-to-day logistics of the campaign for its entirety, making sure things go off without a hitch.

The Expert. If your celebrity is not the technical wizard behind the product or service, make sure that the person who is sitting right next to him is. If you are raising money for a product, someone on your team better be able to be the expert and answer the tough questions about how you're going to make it happen. He or she will understand what you can and can't promise backers and will provide product specs, timelines, and technical explanations, giving credibility to your entire operation. When it's time to actually develop the product, the expert takes over.

Graphic Design Lead. I highly recommend bringing a full-time graphic designer (or design lead) onto the team. The designer will lead the development of all logos, infographics, visual press releases, video animations, project updates, cartoons, emails, T-shirt designs, giveaways, flyers, stickers, and pamphlets. Again, these are all elements that can (and should) be crowdsourced from the sites discussed in the previous chapter, but your design lead will coordinate the content. As I reflect on the ARKYD campaign, two things stand out. First, consistency is key. Our design, look, and feel stayed similar throughout the campaign, helping us scale quickly. Second, sweat the small stuff. Web comic artist Matthew Inman, aka the Oatmeal, became one of Planetary Resources's biggest affiliates. He donated $10,000 of his own money and drove an immense amount of traffic to the campaign because he saw a four-inch ARKYD telescope sticker on a lamppost in the middle of Seattle. You never know who is going to stumble upon your material, get inspired, and join your team.

Technology Manager. Crowdfunding requires a bit of digital dexterity. The technology manager should be part IT guy, part web developer, and part videographer. He or she should be familiar with best practices in technology management. For ARKYD, our technology manager built the Kickstarter website, edited video, set up live streams and Google Hangouts, coordinated audiovisual equipment at live events, and helped integrate solutions across different platforms.

Public Relations Manager (optional). As mentioned, Eric Migicovsky had to hire an external PR team when his campaign went viral and the media came calling.[23] Some projects are focused on niche markets and won't generate this much press. However, if you desire to raise a large amount of money, you need to reach a large number of people. Getting your campaign talked about online, especially when that digital media links back to your campaign page, is especially useful. A PR manager can both help generate attention and—equally critical— help dispel entrepreneurial myths (for example, the fact that everyone building a cool product is certain *Wired* will pick up the story). Having a professional around is going to save you from chasing pie in the sky and help you focus on real ways to move the needle.

Super-Connector (optional). Super-connectors are influential individuals who have access to a vast network of important people, money, and ideas. They usually have large followings themselves and thus know a lot about idea distribution and success. They can help brainstorm marketing strategies for the campaign, internally motivate and inspire the team, implement some of the more ambitious goals, lead behind-closed-door fund-raising efforts, and really build momentum during the campaign. If you know a super-connector or can figure out how to inspire one to help (typically by aligning your campaign goals with theirs), then you will have a huge advantage over campaigns that don't have this access.

6. SHARPEN YOUR AX: PLANNING, MATERIALS, AND RESOURCES

Abraham Lincoln is famous for saying, "Give me six hours to chop down a tree, and I will spend the first four sharpening the ax."[24] The same is true here. Preparation is everything when it comes to crowdfunding.

Planning and Coordination. Crowdfunding campaigns typically have a lot of moving parts, making significant preplanning critical. Go into the campaign with an extremely detailed strategy and logistics map. The campaign manager should keep a master calendar of all meetings, hangouts, events, check-ins, and calls. Team members should have access to this calendar so that everybody stays on the same page. Schedule biweekly or thrice weekly check-ins with members of your team, affiliates, sponsors, and customers. This ensures coordination of your efforts and helps you understand which milestones you will hit and which require pivoting.

Materials. While campaigns vary, some common items you'll need for launch include a prototype or rendering of your product, the campaign video, a crowdfunding platform web page, company or product web page, prewritten emails and announcements, physical promotional materials and handouts, logo and content designs, infographics, and miscellaneous incentives and perks (such as T-shirts and posters). Many of these materials will need to be developed in-house, or via the crowd on websites such as Freelancer, Tongal, or 99Designs. Bottom line: the more you finish before launch, the better off you will be.

Resources. It is easy to underestimate the costs associated with running a crowdfunding campaign, in terms of both time and money. At Planetary Resources, once we pulled the trigger on the campaign, we spent four solid months extensively planning, organizing, and strategizing. Costs incurred during the campaign included advertising (Google, Facebook, Kicktraq, etc.), supplier fees (marketing, creative costs, PR, legal), Kickstarter fees (Amazon hosting, Kickstarter percentage take), physical fulfillment (T-shirts, patches, models, cards, etc.), web applications and education, and contracted work/salaries. As with any digital product launch, you also have to take into account the sometimes substantial costs of faulty payments, refunds, and processing fees.

7. TELLING A MEANINGFUL STORY (AND USING THE RIGHT WORDS)

Traditional fund-raising is something of a niche game. The goal is to please a specific kind of person—a venture capitalist or bank loan officer. Crowdfunding is the opposite. Its focus is exceptionally wide instead of seriously narrow. Every element of a crowdfunding campaign must appeal to the masses. What's the best way to do that? Simple. Use the same technique employed by the very best books, movies, and songs—tell a great story.

The best crowdfunding campaigns draw in backers with powerful, compelling narratives. Consider Let's Build a Goddamn Tesla Museum. The goal here was to buy Tesla's old laboratory and turn it into a museum. But the campaign wasn't about purchasing property or crafting exhibits. Instead, the Oatmeal's comic told Nikola Tesla's story, recounting his brilliant inventions and enormous contributions and revealing the considerable lack of credit that had come his way. And because the Oatmeal is a masterful storyteller, the comic went viral, receiving over 820,000 likes on Facebook and 43,000 mentions on Twitter. People connected with Tesla's story and wanted to help preserve it.

Tips for Telling a Meaningful Story

Make it cohesive. The best tales follow a logical progression. There's a beginning, middle, and end. There are only a few main characters. Confusing potential backers with too much information—too many facts, figures, and spokespeople—does not make for a viral campaign.

Fill a need or desire. In storytelling, never underestimate the power of emotion. Even if the idea seems silly—like, say, a space selfie—if it's deeply compelling and fulfills a basic need, the crowd will listen. People want to be associated with cool stuff, significant events, and inspirational people. Humans make purchasing decisions largely based on emotional impulses.

Focus on the why, not the what. With a product or service, the easiest way to tell a story is to focus on the why. Don't worry so much about explaining what it is and how it works. In other words, remember that the view is different on the inside. If you've been working on a product or service for years, of course all the nitty-gritty details are fascinating to you. But they are perhaps not so fascinating to your audience. Instead, what most people want to hear is why your product/service/idea will improve their life—why it is significant, cool, and important to them and the world. Think solutions and improvements, not explanations or specifications.

Connect with your vertical. Craft your story to target your ideal audience. If your audience is technical, go technical; if they're humanitarian, emphasize the world-changing nature of your solution. However, as mentioned above, even the most technical of ideas needs to be framed inside a greater narrative. If you can't come up with one, tell the story of how and why you came to create the product you're selling. The truth is always the very best story.

Use the right words. In 2014, researchers at Georgia Tech published a study in which they examined over nine million words and phrases used on Kickstarter to determine which language leads to success.[25] The most important lesson is that the words and phrases associated with reciprocity and authority produce the best responses, while projects that focus too much on the need for funds fail.

The most successful language can be broken into the following categories:

- Reciprocity, or the tendency to return a favor after receiving one, as evidenced by phrases such as "also receive two," "pledged will," and "good karma and."
- Scarcity or attachment to something rare, as shown with "option is" and "given the chance."
- Social proof, which suggests that people depend on others for social cues on how to act, as shown by the phrase "has pledged."
- Social identity, or the feeling of belonging to a specific social group. Phrases such as "to build this" and "accessible to the" fit this category.
- Liking, which reflects the fact that people comply with people or products that appeal to them.

- Authority, where people resort to expert opinions for making efficient and quick decisions, as shown by phrases such as "we can afford" and "project will be."

8. CREATING A VIRAL VIDEO: THREE USE CASES, SHAREABILITY, AND HUMANIZATION

Most crowdfunding platforms let you post a short video to help potential backers understand what they are funding and why it is important. This may sound optional, but if you're serious about funding your campaign, it isn't. Fixed-funding campaigns with a pitch video raised 239 percent more money than those without.

Video Tip: Identify three use cases. The best crowdfunding campaign videos target one to three markets. For example, Migicovsky's videos shows footage of the Pebble being utilized by cyclists, runners, and open-source developers. While there are certainly many other ways to use the watch, he focused on three of their largest verticals, keeping things simple and clear.

Video Tip: Put faces to ideas. The video is the perfect way to introduce your team. While it's important to show as many members as possible, the best videos feature one main character—the celebrity—narrating the story and explaining the product. Viewers need a face to associate with the product; having too many faces to keep track of gets confusing.

Video Tip: Show, don't tell. People need to see something to believe in it, but more important, they need to see something to pull out their wallets. If it's a campaign for a new product, then the prototype or rendering of your product has to be a prominent feature in the video. And it shouldn't just sit there. The easiest way to convey the value of your idea to a potential backer is to show people using it. The good news in these days of 3-D printing and computer animation is that it's very easy to create compelling visuals of your product.

Video Tip: Keep it short. Indiegogo found that campaigns with videos under five minutes are 25 percent more likely to reach their goal than anything longer. In 2012, their average campaign video length was 3:27, while the average length for campaigns that reached their goal was 16 seconds shorter (3:11).[26]

Video Tip: Get feedback. "Before we launched Pebble," says Migicovsky, "more than a hundred people had watched our video, seen the page, and given feedback."[27] Not only did it make their video better, but it also validated their ideas and helped shape the focus of the campaign.

9. BUILDING YOUR AUDIENCE: THE THREE As

Having an authentic community of supporters and partners before you launch can be critically important. These people will provide initial momentum and, if managed correctly, will help make your campaign super-credible, jumpstarting the fund-raising process by carrying your information into their networks. For simplicity's sake, let's break down this community into three parts: affiliates, advocates, and activists. Make it a priority to actively build and support these groups in the months before launch.

Affiliates. Affiliate marketing is the practice of partnering with influential individuals, companies, or community organizers to release a product or service. There are two keys to affiliate marketing: picking the right affiliates and designing the right incentives to minimize cost, maximize value, and excite participation. In both cases, the trick is alignment.

Picking the right affiliates. The ideal affiliates share your vision and your customer base. The affiliate's audience must be a coalition of the willing—willing, that is, to do what the affiliate asks when he or she asks. When we started planning for the ARKYD launch at Planetary Resources, we thought science museums would be our best partners, so we went out and built a coalition of five top science centers. We were wrong. As it turned out, science museums tend to have an older audience and a very young audience and neither are particularly Internet savvy or familiar with crowdfunding. Ultimately, we ended up partnering with people like Bill Nye (the science guy, and chairman of the Planetary Society), Brent Spiner (of *Star Trek* fame), Hank Green, Jorge Cham (PHD Comics), Rainn Wilson (the actor), and Matthew Inman (the Oatmeal), all of whom have heavy followings of rabid fans, which is exactly what you're looking for.

Designing the right incentives. In typical product launches, affiliates often take a percentage of sales in return for their help in selling and spreading the product. In crowdfunding, this is too complex and expensive to do effectively. Instead, we came up with creative solutions that excited our partners. For example, we agreed to send one PHD Comic's fan's actual PhD thesis into outer space if the fans shared our campaign with their social community. They loved it. The bottom line: Design programs to spread the word about your project that are *self-promotional* to the partner.

Advocates. Advocates are the fans and supporters of your cause. These are the folks who follow you on social networks, enter their email address to be on your mailing list, and tell their friends about your launch. It's very important to build your mailing list and social following in the months before launch. In the case of Planetary Resources, we put up a splash page on the website asking people to join our mission and give us their name and email.

Eric Migicovsky cultivated his preexisting inPulse fans, leaning on their feedback to help design the Pebble, then turning them into the first wave of crowdfunding supporters. By spreading and sharing the Kickstarter page at launch, these fans helped the campaign go viral.

What to do if you don't have active followers? First, take the time to correctly identify possibilities. Make sure you understand who is going to be interested in your offering and why. Next, find their online hangouts and reach out to them with an invitation to your website. Finally, on your website, have a capture page that invites them to join your community. One of the best ways to get email addresses is to use "ethical bribes"—trades. What can you trade potential customers in exchange for their email address and membership in your community? The simplest answers are often best: an invitation to receive access to your monthly blog, future discounts on products, early access to limited edition products, invitations to events, or in some rare cases (like PayPal back in the day), money. Get creative.

Activists. Activists are those avid supporters who want to do substantial and significant work for the campaign. For Planetary Resources, we created an army of core supporters we called our vanguards. To get this moving, a few months before the campaign, we sent an email to our list of 25,000 names. The email teased:

Hi there—

Everyone on the Planetary Team knows *the moment*. The moment when we knew our calling was to break boundaries and push humanity to the stars! For many of us, that moment was sparked by a mentor, volunteer or educator, often at a science center or museum.

Because of this, educating and inspiring the next generation has been a guiding force for everything we do. This month, we're partnering with a major science organization to create a revolutionary way to make space accessible, interactive and fun. But to make it happen, we NEED YOUR HELP. We're building a team. A select team. We're turning to you, our supporters, to look for a couple hundred members. Only a small number will be selected. We're calling this group the *Planetary Vanguards*. While we can't share all the details, we can tell you that the *Planetary Vanguards* are going to be an important, driving force behind making space accessible in a BIG way! Interested in joining the *Planetary Vanguards*? Here's what you need to do:

1. Fill out the application here. This will allow us to narrow our search for the right people to be a part of the Planetary Vanguards.

2. We'll reach out to you to confirm your interest and verify your contact information. Please note: for some, nondisclosure agreements may be required.

3. We'll invite you to an exclusive, interactive Google Hangout with our cochairman and cofounder Peter H. Diamandis to bring you into the fold and provide you with a confidential briefing.

End of Planetary Transmission.

Chris Lewicki
President & Chief Asteroid Miner

As you can see, we kept the details, names, and dates of the campaign a secret. Thousands of responses poured into our database. We asked them to fill out a questionnaire estimating how much time they could volunteer each week and how big an email list, Google+, Facebook, and Twitter following they had. We filtered these down to about five hundred names, then invited each to an exclusive Google Hangout for a "confidential briefing." Part of this was about engaging our fans, but equally important was that we tested the idea of a crowdfunded telescope on them.

In the months prior to launch, we gave our vanguards assignments, held private meetings, and used them as test subjects for the different strategies we were considering. Then when Chris Lewicki, Eric Anderson, and I launched the campaign with a live event in Seattle, more than fifty of our vanguards showed up in person, some even flying in from Europe. All of them were ready to help. They weren't getting paid. They had traveled on their own dime to be a part of something important.

How critical were these core supporters? Of the 10,000 or so unique clicks we received on the first day of the campaign, almost 50 percent were attributable to the vanguards.

The lesson? Find your most enthusiastic fans and put them to work. They love helping, and their contribution can be invaluable.

10. SUPER-CREDIBLE LAUNCH, EARLY DONOR ENGAGEMENT, AND MEDIA OUTREACH

How you announce your campaign is critically important. The first few days after you launch is when you'll gain the most traction and raise the most money. To come out of the gate with sizzle, there are three key things to remember.

Launching with super-credibility. As we discussed in chapter 5, when you launch above the line of super-credibility, people instantly accept your project as real and believe in their hearts that it is going to work. The key is bringing together as many credible sources as possible and aligning their efforts with yours. Credibility comes from the quality of your video, who is in your video, endorsements on your crowdfunding website, and, if you can muster a launch-day press conference, who is at your press conference.

Super-credibility also comes from the early success of your campaign. People love to back a winner. The better you do out of the gate, the more people will want to support you. This is where the next element comes in: early donor engagement.

Early donor engagement. In an earlier section, I explained that the $5,000 and $10,000 reward levels were as important as the low-priced options. Here's why. In the weeks prior to the ARKYD campaign launch, Eric Anderson, Chris Lewick, and I reached out to our personal networks for help. The day before the campaign went live, we each sent out a couple of dozen emails of the following type:

> Hi Larry,
>
> Tomorrow, Wednesday, May 29, I'm launching one of the biggest and most exciting online projects of my life. No exaggeration. I'd love your support. It's called "ARKYD: A Space Telescope for Everyone." We are crowdfunding an orbiting space observatory (The ARKYD). We're using crowdfunding to generate global awareness and to increase the observation time we can donate to science centers and K-12 schools. This project is about making space accessible!
>
> How this Crowdfunding Campaign performs during the first couple of hours (i.e., pledges received) determines A LOT about its overall success. For this reason, I'm reaching out personally to my close friends—asking for you to consider pledging and passing this along to friends.
>
> There are two levels ($10K and $5K) for you to consider. Both allow you to donate significant telescope time to any school, museum, or university you desire. You can read all the details on campaign page when it launches—it's fair to say that we've put all the top benefits (a lot of them) into these top two levels . . . We'll even name an asteroid we discover after you!
>
> Here's the campaign for your viewing—take a look! But keep it to yourself: www.kickstarter.com/projects/1458134548/1966069095?token=2ab031d1
>
> I will email you a link to the Campaign Page as soon as it goes live. Let me know. Thanks!
>
> Peter

When we launched, the public saw us raise $200,000 in the first four hours. This much momentum helped us establish early credibility and created enough visibility to send the campaign viral. With such a solid funding base, we were unquestionably going to make our target. The only question was when and how much?

Hype. Create meaningful hype. People need a compelling reason to participate in the campaign, and particularly in the launch. In the case of ARKYD, we began teasing our community weeks before the launch, drumming up excitement with "something big is coming" hints. We then used our resources and those of our affiliate partners to promote a live press conference at Seattle's Museum of Flight. We organized ahead of time, ensuring that four hundred enthusiastic fans (from the Seattle area) showed up at the event. For people tuning in via live stream, we offered exclusive perks—a T-shirt—to those pledged while the announcement was going on (about an hour).

Behind the scenes, we worked our network. One common misunderstanding about crowdfunding is that most of your backers are strangers. The truth is that crowdfunding is often a combination of normal fund-raising strategies (seeking capital from your own network) *and* crowdfunding strategies (seeking anonymous public donors).

Engaging the media. The major mistake most people make in crowdfunding is their assumption that simply posting your campaign on Indiegogo and Kickstarter is enough. It's not. You, not the platform, are responsible for driving traffic to your campaign page. And the more traffic you drive to your page, the more money you raise. It's that simple.

In addition to social media and direct outreach via email, affiliates, and advocates, another critical mechanism for driving traffic is digital media—online articles and blogs that link directly to your campaign page. Here are some ideas.

Approach those who know you. If you or your company have been in the media before, create a list of those media outlets familiar with your track record. Contact them and prebrief them. Arm them with a press kit or link to images and content they can easily use.

Create a list of relevant bloggers and journalists. Who are the bloggers who care about your area of interest? The Pebble Watch campaign did this masterfully. "We looked at every single blogger in the gadget blog space and charted how often they wrote about Kickstarter projects and created a list of eighty or so," said Migicovsky. "Next we created a spreadsheet of the top sixty or seventy media journalists that we would look at contacting the moment that our Kickstarter project went live."

Pebble chose Engadget as the exclusive launch media partner. They traded an in-depth news article about Pebble for the right to be the first to break the story. The partnership worked. News outlet after news outlet referenced the Engadget article and the campaign went viral.

11. WEEK-BY-WEEK EXECUTION PLAN—ENGAGE, ENGAGE, ENGAGE

It's critical to stay in touch with backers and would-be backers throughout the campaign. Start before the launch and keep going. According to research conducted by Indiegogo in 2012, projects with regular updates—blog posts, videos, and so forth—raise 218 percent more money than those without.[28] Even better, given the enormous number of communication channels now available and the incredible ease of use, engaging your community has never been easier.

Why is engagement so important? First, backers care about their money. They want project status updates. These are your first customers, so keeping them enthusiastic should be a priority. This is especially true in a fixed-funding campaign, where their contribution is processed only after the fund-raising goal is reached. During the length of a campaign, disgruntled supporters can always lower or even cancel their pledge—keeping them engaged is critical.

Second, engaged backers invite their friends to the party. A huge portion of the capital raised via crowdfunding comes from referrals. And the best referrals come from people who have already contributed to the campaign and continue to be excited about its possibilities.

Third, over 10 percent of the funds raised in our ARKYD campaign came from upsells. This means that during the campaign, backers who had already contributed actually decided to donate more money for a better perk package. These upsells were largely driven by high-engagement activities. Let's take a closer look.

Promotions and contests. People like to play, too. One of the more successful strategies we used for ARKYD was a design contest on Freelancer.com. We partnered with Matt Barrie by putting up a $7,000 prize for the best T-shirt design incorporating the ARKYD space telescope. We were expecting a couple of hundred submissions. Then Matt emailed his list—all eight million of them. Before we knew it, we had over 2,500 high-quality design submissions and an enormous amount of engagement. Plus, the winning T-shirt design became another perk, ultimately driving more sales.

As a side note: When designing promotions, make sure to align the contest with the campaign's ultimate mission. We ran a few contests in communities unfamiliar with crowdfunding and the Internet—they fizzled.

Live-streaming. During the ARKYD launch, we hosted a series of live-streaming Q&As with our celebrities, in which I interviewed Rainn Wilson, Bill Nye, and Brent Spiner. These events helped make the campaign more transparent, promoted deeper engagement, drove people to our campaign page, and ultimately enormously increased our donations.

12. MAKE DATA-DRIVEN DECISIONS AND FINAL TIPS

The details are in the data. As crowdfunding campaigns gain steam, smaller trends and larger patterns start to emerge. Knowing how to leverage this information can give your campaign a huge leg up. Here are a few patterns to look for.

Timing. Timing is everything in crowdfunding. And there are two different timing sequences to pay attention to: market timing and launch timing.

Market timing means the world has to be ready for your solution. Why was the Pebble such a hit? Because the general public was hungry for affordable smartwatches. Be aware of development trends. Pay close attention to increased sales of similar products and do your homework. One of the best ways to test the market is to ask a hundred friends, family, colleagues, and—especially important—complete strangers what they think about your product/service/idea. Do this well before launch.

Launch timing means understanding that people follow schedules. Fewer folks are on their computers during the summer and on weekends. Take into account school holidays, religious practices, and even sports schedules when choosing the best time to launch. Internet traffic is higher earlier in the week. Launching on a Friday, Saturday, or Sunday is terrible from a media standpoint, so if you can't launch on Monday through Thursday, then delay a few days. Don't risk losing early momentum to bad timing. Most crowdfunding ventures have peak activity at the very beginning and very end of the campaign and a lull in the middle. Plan accordingly.

Trend surfing. You want to launch your campaign on a rising tide. Trends matter. Term popularity is important. Check out Google.com/trends. Trend surfing means riding the wave of a trending keyword just as it's becoming viral. Position yourself correctly and you'll surf the wave to its peak, using the term's popularity to drive traffic to your campaign. In our ARKYD launch, this was the idea behind the space selfie. We weren't sold on the idea until we searched for the term *selfie* on Google Trends. Based on the number of global searches we found the term was quickly rising in popularity, so we gave it a shot. That shot certainly paid off.

Upselling. One of the best ways to gain momentum during the campaign is to rally supporters to increase their contributions. Using this foot-in-the-door technique during periods of high traffic can increase sales significantly. But don't beg your backers to buy more; instead, add engagement value at each reward level so backers become a substantially larger part of your movement when they level up.

Global focus. With the Internet, crowdfunding campaigns are no longer local affairs. For ARKYD, we had to translate our campaign page into multiple languages to cater to our international audience. Over 20 percent of our sales came from European countries. Three percent of our sales came from China. Pebble's 68,000 backers came from all over the Internet and all over the world. "Our news spread from North America to Canada to Europe to Belgium to Holland to the Middle East to Singapore to Indonesia to China to Japan," says Migicovsky. "Overall, the stats showed that roughly 50 percent of our backers came from North America and 50 percent came from elsewhere in the world. [And] it wasn't just the English-speaking world."[29] In other words, do your research and understand the global market. Prepare your campaign accordingly.

Ask questions and listen to your community: Campaigns are not static. This isn't launch and forget; it's launch and get busy. And get busy with your data. Gauge the opinions of your customers constantly. Make iterative improvements based on the information collected. Making data-driven decisions throughout the campaign can dramatically improve your chances of success. Here are a few pointers:

Listening to your community will not only help fund your campaign but will also provide important feedback about the product/ service you're offering and the right incentive structure to use. As you collect data, try to understand two things: What do people want? How much are they willing to pay for it? As expected, surveys work well in this context. Sometimes simply asking your customers and community exactly what you want to know is often the best approach. So remember to:

1. *Segment the audience.* By reaching out to specific groups of
   people within the community, you can draw better conclusions.
2. *Ask only one question.* People are busy. Answering one question is easy and doesn't take too long. Ask the right question.
3. *Expect exaggeration.* Be aware that people tend to choose
   the extremes in surveys. When it comes time for them to actually contribute, only a small percentage will actually put up the
   amount they selected in a survey.

So that sums it up. There's a lot more information online at the www.
AbundanceHub.com site. My final piece of advice comes from our
friends at Nike: Just do it. A major crowdfunding wave is just now hitting, and it will be growing tenfold over the next seven years. Don't
miss it. Pick a project, a product, or a service and get busy building
your campaign.

Hold the Presses—Some Final Advice

Shortly after submitting this manuscript to my publisher, a Portland-
based product development whiz named Ryan Grepper crushed all
previous crowdfunding records—including both Pebble Watch and
Ubuntu—getting 62,642 people to pledge $13,285,226 to support
his Coolest Cooler campaign. What were they supporting? The creation of a beach cooler for the twenty-first century—a cooler that
comes with a built-in blender, phone charger, Bluetooth speakers,
and, of course, a waterproof lighting system so you can find your bev-
erage of choice after dark. Since my close friend, the marketing genius
Brendon Burchard, had helped with this campaign, I reached out to
him to see how they pulled it off. Turns out, during their campaign,
Brendon and Ryan had uncovered four ideas not discussed in any
of our previous examples, but with enough importance that a last-
minute insertion was justified. So here's a quick look at what they
learned along the way.

1. *Fail Forward.* Despite their enormous success, the Coolest
   Cooler wasn't an overnight sensation. In fact, this $13 million
   dollar success was Ryan's second effort at crowdfunding the
   cooler; his first failed to raise $125,000, and the story almost
   ended there.

   And it's a long story. "Over a decade ago," explains Ryan, "I
   made a beach blender out of an old Weedwacker. It made for
   fantastic family outings. The next year, I took my old car stereo
   and added it to a cooler so we could also have music on those
   trips. For years, these gadgets worked great, but then we moved
   and I put them in storage and didn't think much about them.
   But, last year, I pulled both the cooler and the blender out of
   mothballs and realized that technology had come a long way
   since I'd built them. Today, I could combine much more technology into a smaller, more portable cooler. And that's when it
   hit me—this could make for a great crowdfunding project."

   But not so fast. Ryan's first attempt at crowdfunding the
   Coolest Cooler actually failed. In a late 2013 effort, he only
   raised $100,000 of the $125,000 needed to reach his funding
   goal. Disappointed, but not deterred, and urged on by supporters from his first effort, Ryan took the cooler back to the
   drawing board—adapting, iterating, and ultimately succeeding.
   And this brings us to our first lesson: Fail forward. "The fact
   that your first crowdfunding effort failed doesn't mean you can't
   relaunch," says Ryan. "In general, crowdfunding is a great way
   for any creative to put an idea in front of potential customers
   and get the most real and honest feedback you can imagine—
   feedback from their wallets. If you fund, fantastic. If you fail,
   you say, You know what, this wasn't as good as I thought it was.
   So what can I change, improve, and iterate? Or, 'fail early, fail
   often, fail forward.'"

2. *Start with a Crowd.* The second lesson Ryan learned was the
   importance of having a community of people who care about

your product. Earlier in this chapter, I showed you how to create a group of Vanguards to play this role. In the case of the Coolest Cooler, Ryan's supporters from his first campaign filled that bill. "During that campaign," he explains, "we did a lot of outreach—going to tailgating parties, networking, connecting. It was these same people who encouraged me to try again. And when I did, during our second campaign, these preexisting fans were with us from the start. They were a big boulder we threw into the pond. It made a very big ripple and had a very big effect."

3. *Mockup Matters.* The Internet is a visual medium. It's why people care so much about the size of their screens. It's also the reason that we talked about the importance of having a great video for your campaign. What Ryan learned is an extension of these facts—that having a great mockup of your product to display in that video is equally important. "The extra time between campaigns allowed us to build a great quality mockup," he says. "It made a big difference. And it's not all about looks—it's actually about trust. With crowdfunding, you're asking strangers to care and to believe in both you and your concept. You're asking for a big leap of faith. Showing them how close you are to having an actual production model—something both finished and great looking—provides an incredible leg up."

4. *Targeted Advertising.* Market segmentation matters. "It's critical to craft your message for a specific audience," explains Ryan. "From our first failure, we had some data about our backers—who they were and what kinds of things they were excited about. So we put together targeted Facebook ads and connected with people who already had passion for boating and tailgating and camping and picnics and Jimmy Buffet and such. When you understand your buyer, the targeting power for some of these advertising platforms is astounding." So here, our fourth lesson

is the value of paid targeted advertising through platforms like Facebook, YouTube, and LinkedIn. Be a data-driven company. If the ads help your campaign raise more money than you're spending on advertising, then go for it.

Finally, I'll add one last point I found fascinating: The campaign's incredible level of success brought an entirely different type of benefit. Ryan recalls, "One guy reached out saying, 'Do you need $20 million in financing? I've got a team, and we can make this the next $200 million company.' On top of that, we had countless partnership proposals for almost every major component maker—from batteries to blenders to speakers. Even potential distributors got in touch."

CHAPTER NINE

BUILDING COMMUNITIES

Reputation Economics

"The trillions of hours of free time the population of the planet has to spend doing the things they care about" is how NYU professor Clay Shirky defines the term *cognitive surplus*.[1] The entire third part of this book has been spent exploring strategies for tapping this surplus, but we've come to one of the most important lessons: how to build a community that you can tap into, work with, and use to accomplish the big and bold.

Let's begin with what we mean by *community*. For starters, it's different than the crowd. The crowd is everyone online. A community is pulled from the crowd. It's everyone with whom you have a working relationship. There are different types of communities, but in this chapter we're going to focus on two: DIY communities and exponential communities. A DIY community is a group of people united around a massively transformative purpose (MTP),[2] a collection of the passionate willing to donate their time and their minds to projects they truly believe in. These folks work for free. They work long hours. They remain committed. And they do so because they feel the work is meaningful and important. An exponential community, meanwhile, is a group of people who are immensely passionate about a particular exponential technology (machine learning, 3-D printing, synthetic biology), and who unite to share techniques and experiences.

On a certain level, there's nothing new going on here. As long as humans have lived communally, they've been banding together to share passions and tackle problems. But today's DIY and exponential communities (which I will refer jointly to as communities for the remainder of the chapter) are distinguished by differences of geography, scale, and structure. Let's take a closer look.

Geography is the most obvious change. Historically, if you were building a boat and the best maker of masts was located on the other side of a mountain range—well, you were out of luck. But the Internet removes these barriers, and does so in important ways. As Bill Joy famously pointed out, the smartest people on the planet usually work for someone else.[3] Technology now gives you access to these big brains no matter where they live and without a host of traditional biases. Online, no one can see what color you are, which gods you worship, how you dress, what your hair looks like, if you smoke or smell or smile too much. This anonymity allows people who wouldn't normally sit on a park bench together to share deeply meaningful and potentially profitable experiences.

Moreover, liberation from proximity and prejudice increases access to new ideas. Since creativity is recombinatory—i.e., breakthroughs result from new ideas bumping into old thoughts to produce novel insights—this increased access to ideas amplifies the rate of innovation in communities. In fact, if you combine this amplified rate of innovation with our newfound ability to tap any expert anywhere in the world, the potency of technologies like 3-D printing and cloud computing, and the power of crowdfunding to capitalize such ventures, you find the second key difference in today's communities: the scale of projects they can now undertake has grown exponentially.

Communities are now empowered to tackle jobs far larger in scope and size than anything previously possible. For one example, the online hobbyist community DIY Drones has been able to build

military-grade autonomous aircraft; for another, Local Motors is constructing fully customizable automobiles.[4] Ten years ago, challenges of
this size were the sole province of large corporations and governments.
Today they're open to anyone with access to the Internet.

Structurally, the change has been equally significant. Consider that
the major structures that handled the dissemination of information of
the last century—radio, television, print—went in only one direction.
Communication flowed from the top down, and even then just barely,
by contemporary standards. But today the Internet permits top-down
and bottom-up and side-to-side communication. In communities,
these new possibilities for communication don't just allow a leader
to lead, they allow other leaders to emerge. They permit collaborative
structures that would have been unthinkable just a decade back.

Even better, in many cases, these structures are self-organizing. If
the community has been set up in the right way, then growth happens
organically, without need for too much direct intervention or intensive
capital spends. For example, after Facebook COO Sheryl Sandberg
wrote *Lean In*, her bestselling book on empowering women to pursue
their ambitions, she decided to capture the energy it was generating
by building an online women's community. As part of their growth
strategy, one of their ideas was to create Lean In circles—local groups
of eight to ten women coming together to share experiences and offer
support. "These circles are almost entirely self-organizing," says Gina
Bianchini, CEO of Mightybell, the online community building tool
that serves as the backbone for the Lean In community (more on such
platforms in the how-to section). "All they did was suggest the idea to
the community and issue a set of loose guidelines for how these circles
should form and function. There is no one in the Lean In organization
whose job it is to create new circles or manage existing ones. But in
a little over a year, some 13,000 circles have been formed, with more
starting up every week."[5]

The driving force behind all of these novel collaborative structures, self-organizing or otherwise, is an entirely new kind of value
proposition, what technology expert and author Joshua Klein calls

reputation economics.[6] The idea here is twofold. It starts with the fact that some two billion of us now have online reputations. Whether it's your seller rating on eBay, or the content on your Facebook page, or your Klout score (Klout uses social media analytics to rank its users according to online social influence), people know far more about each other than ever before. And these reputations matter. A series of powerful blog posts can get you everything from a date for Friday night to an invitation to speak at a conference. People get jobs because of their StackOverflow.com experience (a website that lets techies comment on one another's questions and vote up the best answers) and TopCoder scores (TopCoder runs online computer programming competitions). In other words, our online reputations have real-world consequences.

Moreover, these reputations allow all sorts of entirely beneficial but—the key point—not always financial exchanges. "Because we can now get context-relevant information about anyone else in the world," explains Joshua Klein, "we can decide, dynamically and personally, how to exchange with them in a way most beneficial to both parties. Essentially, this underpins everything that makes every community with an online component work, which these days is most of them."[7]

But the interesting bit is that we can now take this one step further. Imagine a local baking club that's getting into carbon dioxide infusions for their whipped-cream cupcakes. One day one of the club members decides they need a more industrial-strength infusion machine. A few minutes later she's looking at the website of a guy halfway across the country who claims to be making a device that perfectly fits the bill— but he's got only a couple of prototypes and doesn't seem interested in sharing.

"Used to be," continues Klein, "you'd try to figure out what he was doing from papers he may have written or a newspaper report, or maybe you'd write a letter and beg for more info. But now you can learn all about this guy. Figure out he's really into Bavarian folk music. Bavarian folk music? No kidding, your lead baker's wife's mother's aunt is from Bavaria. Turns out her nephew back home is a big deal in folk music. Would this guy making industrial-scale cupcake infusers be interested in an introduction, maybe for a trial of his device? That sort of thing just wouldn't have been possible ten years ago. Now it's as ordinary as breathing."

By fundamentally altering the value proposition, reputation economics further accelerates the rate of innovation both in DIY and exponential communities. This means that communities don't grind to a halt when money is not readily available. In fact, often just the opposite is true. Mutually beneficial nonfinancial trades can actually be better—that is to say, add more value—for the participants involved than a plain old currency exchange. There's less friction, so people are often more motivated to make such trades. As a result, the accelerated rate of innovation that results from the removal of geographic barriers is itself accelerated, allowing entrepreneurs to go from A to B far faster than ever before.

Or as Gina Bianchini explains: "I've been around DIY communities my entire career, and these continue to surprise me, to startle me. Once a community gets going, it starts generating absolutely mind-blowing ideas of its own. Directions you would have never considered, never even imagined. This happens so regularly you can almost count on it. I think this is the reason DIY communities are such a powerful tool for tackling bold challenges. You can go big because you don't need to know how to pull something off ahead of time. The community shapes the path and accelerates the process. It's a shocking amount of leverage."[8]

Case Study 1: Galaxy Zoo—A DIY Community

In early 2007, while working toward his PhD in astrophysics at Oxford, Kevin Schawinski was hunting blue ellipticals in the Sloan Digital Sky Survey data. A blue elliptical is a transitional galaxy, possibly the missing link between a galaxy engaged in active star formation and one long dead. The Sloan Survey, meanwhile, is one of the more ambitious endeavors in the history of astronomy. Imaging almost a quarter of the sky and containing ten times the data of any previous effort, Sloan's goal was to give us a large-scale view of the universe—what the *New York Times* once called "a census of the heavens"[9]—in the hopes of revealing its structural framework. Thus, unlike pre-Sloan astronomers, Schawinski didn't have just a few thousand photographs to work with—he had nearly a million. Unfortunately, at the time, our best computer algorithms couldn't spot blue ellipticals. Only the human eye was capable of that feat. In other words, Schawinski had his work cut out for him.

Ten hours a day for five days straight is what it took him to sort through 50,000 images, but that was the end of the line. "It was as far as I was willing to go," Schawinski explains. "And we extracted some really interesting science and published a bunch of papers from what I had analyzed, but whenever we talked about what else we could do we came back to 'wouldn't it be amazing to sort the whole million.'"[10]

Then Schawinski went out for a beer with fellow Oxford astronomer Chris Lintott and together they stumbled upon the idea of putting the images on a website. "We thought that there might be a few people out there—like, maybe, two or three really dedicated amateur astronomers—who would be willing to help us," says Schawinski. "With this method, by doing a back-of-the-envelope calculation, we thought it would take about five years for every one of those million galaxies to be classified once."

With the help of some friends, and just two weeks after they shared that beer, this idea became Galaxy Zoo, one of the very first citizen-science websites to appear online.[11] As far as telling people about the website, well, they commemorated its appearance with nothing more than a short press release. Then they waited—but not for long.

Within a few hours, people were classifying more galaxies than Schawinski had done in a week. Within twenty-four hours, they were classifying nearly 70,000 galaxies an hour. "What we realized very quickly was there was this huge demand among people to get involved in this. At first we were kind of puzzled by why people would want to go to a website and classify galaxies in such huge numbers. Then we realized we'd hit upon an unmet need—people want to do this. They wanted to contribute. In fact, we teamed up with some social scientists and found that the number one reason why people do Galaxy Zoo is the desire to contribute to actual science. They want to do something that's useful."

A lot of people had this want. Schawinski and his colleagues had hit on a massively transformative purpose. The first iteration of Galaxy Zoo (they're now up to version five) drew 150,000 participants classifying—wait for it—50 million galaxies. Subsequent versions pulled in over 250,000 participants and pushed the total over 60 million. And then Galaxy Zoo became a smorgasbord of citizen-science projects, now hosted at Zooniverse. Want to explore the surface of the Moon? Join Planetary Resources and NASA to find near-Earth-approaching asteroids for potential mining? Model climate change through the centuries using historic ship's logs? Help researchers understand whale communications? All of these choices are now in the offing.

And that's important. What Schawinski and his cohorts had accidentally stumbled upon is what I call the Law of Niches, the idea, quite simply, that you are not alone. This is one of the most telling features of the web—the somewhat humbling fact that no matter what oddball notion you're deeply passionate about, well, there are plenty of folks who share the same passion. "The ability for entrepreneurs to nimbly find and serve niche interests—and to produce platforms that allow those groups to address their needs en masse—is better than ever before," explains Joshua Klein. "It used to be that start-ups would have to compete with an established industry vertical—say, automotive parts. But I've got a friend who is building his entire business around Prius owners who want to hack their cars' electrical system to make them even more fuel efficient. That's a pretty small subculture, but today it's more than enough to build a business upon."

Case Study 2: Local Motors—A DIY Community

John "Jay" Rogers grew up loving cars. He also loved motorcycles. This was something of a family trait. His grandfather, Ralph Rogers, was the last owner of the legendary Indian Motorcycle company and the first distributor of the Cummins Engine on the East Coast. Growing up, Rogers always assumed he'd pursue a career in automotive design, but when he got to college, he discovered there was no place in the traditional university system for car designers, so he set aside his childhood passion and graduated from Princeton with a degree in international affairs and public policy (and a minor in art).

Rogers took a job with a medical start-up, spending three years in China before switching to a career as a financial analyst. That career came with an offer to go to business school, and he was accepted at Stanford. During a celebration dinner, a colleague asked Rogers what he really wanted to do with his life. "I told him I wanted to build something tangible," says Rogers, "to actually lead people. My friend asked me if I knew how to lead, if I actually have any real leadership experience. When I said no, he suggested I join the military."[12]

Which is exactly what Rogers did.

At age twenty-six, he gave up his position at Stanford and became a Marine, signing up in 1999 and serving six and half years, including a tour in the Pacific and another in Iraq. In 2004, on his second tour, Rogers brought along a copy of *Winning the Oil Endgame*, visionary environmentalist Amory Lovins's book about how society can wean itself from fossil fuel dependence. The book was a turning point. He read it right around the time two of his closest friends were killed in combat. The combination made him realize that what he really wanted to do with his life was ensure that no one else ever died for oil. Since 71 percent of the fossil fuels imported by America becomes the gasoline that powers our cars and light trucks, he figured that the best way he could accomplish his goal was to build an entirely new kind of environmentally friendly car.

Rogers knew he needed more business savvy than he had to pull off this dream, so he left the military and went back to school for an MBA at Harvard. It was there he heard a presentation on Threadless, the previously mentioned open-source T-shirt company. He was stunned by the power of crowdsourcing. Certainly, building cars was far more difficult than designing T-shirts, but Rogers also knew that the talent he needed was readily available. In another example of the Law of Niches, Rogers realized he wasn't the only kid who grew up fantasizing about designing cars only to later realize this was a very rare job. "Only 12 to 20 percent of the industrial designers who specialize in transportation end up working in the field," Rogers explains. "And that doesn't include all the people who wanted to build cars but didn't become industrial designers. Or couldn't become industrial designers. There is this huge pent-up need in people to create cars, this very frustrated passion."

The result of this MTP became Local Motors, the world's first open-sourced car company to reach production.[13] Able to design and build cars five times faster and with a hundred times less capital than traditional manufacturing companies, Local Motors is something of a modern wonder. Not only did they figure out how to accelerate and demonetize automotive production, they did so at a time when unemployment in Detroit—thanks to the slow death of the American auto industry—was hovering at 23 percent.[14]

Think about this for a moment. Throughout the book, we've been talking about how small teams can now accomplish what once only large corporations or governments could do. Well, if you exclude Elon Musk's Tesla, America hasn't seen a new car company succeed in thirty plus years. And in the past seven years, the government has spent tens of billions bailing out the Big Three. In other words, Local Motors isn't just doing what large companies and government could do; they're doing what these institutions could not—helping to save the automotive industry.

So what did they do? Simple. Local Motors figured out how to design and build cars collectively, through an incredibly robust DIY community. Today they host design competitions on their website, targeting very specific regional markets (off-road vehicles for the Sonoran Desert, incredibly fuel-efficient vehicles for California's high-traffic freeways). The contests aggregate car concepts from a worldwide assortment of designers, engineers, and enthusiasts. Then the community votes up their favorites, and Local Motors helps brings the winning car into existence.

Critically, Local Motors keeps their community involved at every step. After that first design contest, Local Motors organizes additional design/build competitions around vents and interiors and other key features. They also leverage mass production, allowing their community to vote up their favorite off-the-shelf parts for inclusion in the vehicle. For example, the first car Local Motors released, the 2009 Rally Fighter—an off-road (yet street legal) desert racer—has Mazda Miata door handles and Honda Civic taillights. The company then releases the final design under a Creative Commons license so community members can continue to enhance the work and, for those entrepreneurially inclined, develop specialized parts to sell to the community. Lastly, to take possession of a car, customers must actually participate in the assembly process, cobuilding the finished product at a Local Motors build facility, aka a microfactory.

Of course, what we're talking about here is *engagement*, but that word is often misconstrued. "Look," explains Gina Bianchini, "engagement isn't a like on Facebook. A like is just one-way communication. It doesn't go anywhere. You have to think about what a community actually is—it's people talking to one another. Engagement is always about getting that conversation going and keeping it going."[15]

Local Motors does just that by providing a meaningful outlet for the underserved pent-up creativity of car enthusiasts at every step in their process. They're not just letting their members peek behind the curtain of automotive design—they're helping them become the wizard. As a result, because they are turning loose so much fundamental passion, they didn't even need a huge community to become, as Chris Anderson wrote in *Wired*, "the future of American manufacturing."[16]

And he wasn't kidding. Back in 2013, Rogers and GE CEO Jeff Immelt teamed up to cocreate a Local Motors–style microfactory that specializes in speeding the time from mind to market for GE appliances. That factory opened in 2014. At the time of the ribbon-cutting ceremony, it already had two GE appliances in production. On the heels of that success, Local Motors is building an additional fifty microfactories to drive innovation in other industries.[17]

There is the mistaken impression that to really tackle a bold challenge via a DIY community, the community needs to be huge in order to match the scale of the task. Not true. While today's Local Motors community is over 130,000 active members strong—with another 1,000,000 or so lurkers (those who watch but have not yet participated)—the collective that built that first Rally Fighter was a meager 500 people by comparison. The point is this: If your community can provide a legitimate release valve for people's incredibly frustrated passion, you are unleashing one of the most potent forces in the history of the world.

Case Study 3: TopCoder—An Exponential Community

It started in the late 1990s. Jack Hughes was running a software development company in Connecticut. During those quiet periods between projects, Hughes kept his employees busy by holding in-house programming competitions. A few years later, after Hughes sold that company and was looking around for what to do next, he realized those competitions were something to build upon.

"I was sitting with my brother at a picnic table," he explains, "discussing that fundamental business problem: How do you find really qualified people? A great developer, a great creative person, is a very difficult thing to find. But I already knew—because we had run those internal programming competitions—that developers loved to compete and that those competitions were a good way to identify top talent. Since so much of development work had become web work, I started wondering what would happen if we put those competitions online."[18]

In 2002, Hughes stopped wondering. He and his brother set up the website TopCoder and began holding contests. "Initially we put some money up as prizes, just to keep things interesting, but mostly the contests were for pride."

Pride, as it turns out, was the secret sauce.

Hughes was a longtime sports fan. He loved the bracket-based tournament structure used by the National Collegiate Athletic Association (NCAA) during March Madness. He also loved how baseball analysts had begun to depend on a far wider assortment of statistics than just home runs and RBIs. So Hughes created a leaderboard for TopCoder that works much the same as a NCAA tournament bracket. He also started posting electronic "baseball cards" for each programmer, the box scores displaying everything from statistics on how many competitions a given coder has entered to their highest and lowest scores in those competitions.[19] This rating system was designed so coders could look at the names and ratings of the other contestants entered in a contest and decide if they had a shot at winning and if the contest was worth entering. But it quickly became a point of pride.

"Really," explains Hughes, "we turned coding into a massive multiplayer game. We would post a problem statement, and as soon as a coder opened it, a clock would start ticking down. People got points for how quickly they submitted a solution and how accurate their code was. But how high your rating was, that was a badge of honor. People weren't competing for the money, they wanted the rating."

But the money didn't hurt, so rather than just making up competitions for his community to solve, Hughes solicited outside business. By atomizing big projects into bits, then organizing competitions around each bit, specialists within the TopCoder community deliver solutions one puzzle piece at a time. Some people are great at spotting bugs in code, while some are great at fixing bugs—that sort of thing. After these individual contests have been won, the whole project is reassembled and delivered to the client. It's a great crowdsourcing model.

And it's had enormous impact. In the beginning, a TopCoder community of roughly 25,000 was solving serious problems for the likes of GEICO and Best Buy, but it didn't take long for those numbers to grow and for that community to become involved in efforts significantly farther afield. For example, when Dr. Ramy Arnaout of Boston's Beth Israel Deaconess Medical Center was trying to sift through a huge pile of genetic information about the immune system, he decided that rather than just consult with fellow scientists, he might want to give TopCoder a shot.

"The result," wrote Carolyn Johnson of the *Boston Globe*, "a deeply biological problem—analyzing the genes that produce proteins involved in the immune system's ability to identify microbes—could be rapidly and efficiently answered by a community of more than 400,000 computer programmers."[20]

What's more amazing is how little time and money it actually took for the community to do this work. A two-week contest brought in over a hundred entries from coders in nearly seventy countries. Sixteen of those entries outperformed the algorithm then used by the National Institutes of Health. And—for those interested in using TopCoder to crowdsource programming needs (rather than as a guide to building a DIY community)—the whole contest cost just $6,000.

But if you are interested in looking at TopCoder for exponential community building purposes, the most important thing to remember is that it's not just competition that drove their growth. "Competition just happened to be the thing that first engaged our community," says Hughes. "Many other aspects of TopCoder are about collaboration. People get involved not because of the money or the sponsors or the fact that they can get jobs as a result. They get involved because it's social. We've given our community a place to get together because they want to get together—that's why it works."

In late 2013, Appirio, the cloud services consultancy, purchased

TopCoder and the community was put under the leadership of Appirio cofounder Narinder Singh. "We saw the amazing successes possible through TopCoder," explains Singh, "but we also noted that it was being used by a relatively small number of customers, under limited circumstances. The more innovation a company wants, the greater the access to powerful technology they will need. Our goal in acquiring TopCoder is to make it mainstream technology for small, medium, and large organizations, to give them that 'show me' moment they need to make the exponential capabilities like TopCoder a core part of their technology arsenal."[21]

Who Should Build a DIY Community?

Recall the Law of Niches. The ego-belittling truth the Internet makes visible is that none of us is as unique as we'd like to believe. And this is good news. It means that if you are passionate about something, there's a pretty good chance others share that passion. So the best reason to start a DIY community is unrequited love.

Look, if you can leverage an existing community to fulfill your dreams, go that route. But if you're passionate about something and no one else is sating that desire, then you have first-mover advantage. Don't underestimate this power. When Galaxy Zoo first started, they were pretty sure only a handful of people would sign up to catalogue galaxies—yet within a very short time, tens of thousands of people were involved. Why? There was a deep, unmet need in people to participate in astronomy, and Galaxy Zoo was the only game in town.

We saw something similar with Asteroid Zoo—the Zooniverse-hosted collaboration between my company, Planetary Resources, and NASA to use humans to identify new, never-before-detected asteroids, which, in turn, will create a rigorous dataset from which we can train up AIs to do the same at scale. This is such a specific desire that we were not certain how people would react, but just as with Galaxy Zoo, the crowd wildly exceeded our expectations. In the first six days of the project, we saw more than one million images reviewed and more than 400,000 asteroids classified.

There's a corollary here: If you don't have first-mover advantage, then ask yourself what new and exciting twist you are bringing to the table. Think about CrossFit. The health and fitness space was exceptionally crowded when this workout craze was introduced, but Cross-Fit leveraged two facts: People work harder when they're in a group of peers, and outside of yoga, there were no fitness classes aimed directly at men. So while CrossFit lacked a first-mover advantage, their distinguishing elements were new (no other game in town) and exciting (you get a better workout), and that was more than enough to build upon.

It's also important to remember that people join communities because it reinforces their sense of identity (see below), but they stay for the conversation. This is also why the very best communities actually force their members to interact with one another—they actually drive that conversation. And if you're the person organizing a new community,[22] then driving a high level of interactivity must be your primary responsibility. Every community manager is first and foremost a conversational caretaker. When Chris Anderson initially created DIY Drones, he spent an average of three to four hours each day curating his community. Simply put, if you're not an especially social person or are too busy to put in that kind of time, community building is probably not for you.

Why *Not* to Build a DIY Community

A great many people try to build communities for the wrong reasons, which is no different from building a house on a rotten foundation. Do the latter, and no matter how fancy your door knockers are, the structure will eventually collapse. The same holds true for communities. So before we even get into why you should take the time to build an online community, let's start with the three main reasons why you should *not* build a community.

1. *Greed.* Online communities are about achieving an MTP, not the cash. This is not to say you can't monetize these communities, but this won't happen right away. Communities are about authenticity and transparency, and you need to prove that you're the real deal before you begin asking for real dollars.

2. *Fame.* Reputation economics tells us that one of the main reasons people join online communities is because they want recognition. And the purpose of your community is to give it to them, not to you.

3. *Short-term desires.* For starters, getting a vibrant DIY community up and running isn't going to happen overnight. Moreover, people are attracted to big visions. When Planetary Resources attracted our Vanguards (essentially the foundation of our community), it wasn't because they were excited to help us build a new telescope. They were excited about opening the space frontier.

Stages of Community Building

If you've decided you want to start a community, there are nine key stages you will need to pass through along the way.

1. Identity—What Is Your MTP?
2. Designing Your Community Portal
3. Community-Building Resources
4. Early Days of Building Your Community
5. Creating Community Content
6. Engagement and Engagement Strategies
7. Managing Your Community
8. Driving Growth
9. Monetization

Identity—What Is Your MTP?

People join DIY communities because it reinforces their sense of identity. So start by finding your people. Who is your tribe? What is your MTP? "Passion is the differentiator," says Jay Rogers from Local Motors. "So throw up a flag and be very clear what you stand for." Be as specific as possible. Write up a mission statement, post it in a visible place on your website, then cling to that statement. Put differently, the Law of Niches works only if you very clearly identify and authentically support that niche.

This is also why it's important to tell your story. A mission statement is nice, but unless you're already a public figure, then people need to know who you are and why you're doing what you're doing. This could be in written form, but better yet, turn on your cell phone camera and record a video. Don't forget to let your passion shine thru.

Designing Your Community Portal

There are six basics regarding designing the look, feel, and engagement of your community portal:

1. **Just get started—design something authentic.** LinkedIn founder Reid Hoffman famously said, "If an entrepreneur isn't embarrassed by the first release of their product offering, then they launched it too late." The same applies to communities. You want to start by starting. Don't spend years designing the right portal, and don't blow your bank either. Authenticity matters. And having a personality is key (so people can quickly figure out if they belong in your tribe), but giving people a place to have a conversation is more important. You can always add better window treatments later.

2. ***Navigation.*** People need to know how to move around easily. They want to know where to go and what they'll find there, and if you can't tell them this information quickly and clearly, there are plenty of other places to visit online. In other words, the navigation bar is not the place to get creative. Take DIY Drones, an online community of UAV enthusiasts. The first thing you see when you land on the site is a big box that reads: "Welcome to DIY Drones" and directions for how to use the site, including the ever-important "I'm new to this—where do I start?" tab, prominently displayed.

3. ***Simple Registration.*** If it takes more than thirty seconds to become a member of your community, you're not going to have many members. Similarly, if you want much more than my email address, then I suspect you're secretly planning to make money selling my personal information and I'm not interested. Ask for my email. Tell me exactly what I'm getting in return. Promise me that you're not going to resell my data. And give me an easy way to invite my friends to join as well.

4. ***The Information.*** What to post on your platform is much a matter of personal preference, but it's helpful to remember that people join online communities for four main reasons: a sense of belonging, a support network, greater influence, and a way to sate curiosity/explore new ideas. Most everything you choose to put on the site should be designed to meet these needs.

5. ***Recognition.*** Whether you create a leaderboard/rating system or make your blog open (anyone can post), be sure to highlight popular content (for example, the right side of the DIY Drone's is devoted to Top Content)—specifically, a short description of the blog post and, more important, a larger picture of the person posting. Remember, this is reputation economics; people want to be celebrated for their contributions.

6. ***Scalability.*** Sure, you might think you want a gargantuan membership, but understand that good communities are messy places. This is key. You want some of this mess because it will

generate more new ideas and help accelerate the rate of innovation and make those members who hate top-down authority feel more comfortable, but you also need to be able to steer (not control) the mess. This means you have to give members a way to break into smaller groups. This is why, for example, Facebook is not always the best home for a DIY or exponential community.

Early Days of Building Your Community

No way around it, getting out of the gate is always tough. But you don't actually need many members to have an impact. In fact, as Richard Millington, founder of the community consultancy FeverBee, wrote in his blog: "The bigger a community gets, the less people participate. This creates wastage and makes it impossible for the community manager to identify and work with the top members. Better to extract 1 hour a day from 100 committed members than have 50,000 mostly inactive lurkers. Stay small and extract maximum value from the few, not a little from the many."[23]

So how many members do you actually need? Again, less than you probably think. Most experts recommend handpicking (see below) your first ten to fifteen members so that when visitors drop by there's something interesting going on. Gina Bianchini, CEO of Mightybell, has found that 150 members is usually the point at which the community itself begins to carry the conversation.

Here's how to get started.

1. *Be the First Mover.* It seems obvious, but being the first one into any space gives you considerable leverage. If people want to have a conversation and your community is the only place to have it, you're already winning. If you can't be first mover in a space, then the problem you're here to tackle (your MTP) better be significantly different and arguably more visionary than the competition's.

2. **Handpick Early Members.** Research shows your early adopters tend to become your most ardent supporters. Get the ball rolling by personally handpicking your first ten to fifteen members. Be sure to engage these folks in the community-building process. Ask for their advice. Integrate their input. Don't waste your time going after big names. As a general rule, these folks are busy with their own communities.

3. **Establish a Newcomer's Ritual.** You want to give members a way to feel like they belong—but they have to earn it. Create a ritual and tie it to a specific membership milestone. After a new member has their fifth blog post or one of their comments draws ten likes on Facebook, reward that level of participation with a token.

4. **Listen.** No matter what your core vision is, you can't get anywhere without your members. So pay attention to what they have to say and be prepared to change direction when necessary.

Creating Community Content

There is no way around the fact that running an online community puts you into the content business. While there are plenty of experts who feel that too much engagement from a community manager is not a good thing, too little is an easy way to increase member drift. Most of the founders we spoke with claimed that they were on the site and taking care of their community constantly, especially over the first six months after launch. In other words, they became content production machines. This goes with the territory. Expect it, plan for it, and execute.

Here's a list of the five basic content categories to draw upon, and below that, a chart breaking down how typical communities spread out this content.

1. **The Future.** This takes many forms. You can preview an upcoming event or preview an upcoming product launch or preview

the upcoming week (what will be happening on the site over the next seven days) or make predictions about the coming year. Previews are a great way to keep the community informed, while predictions are a great way to start a debate. Both are useful.

2. **The News.** This can be a news roundup or breaking news or news about just-released products (product reviews). All are commonly used and fairly effective. That said, because plenty of other sites go this route, be sure to find a way to make your news new. Give it an edge. Be funny. More critically, be sure to do a member news section. What is the community doing? Did someone do something amazing or change jobs or bump into a VIP? Using your site as a way to celebrate member achievements is a great way to foster loyalty and enthusiasm.

3. **The Interview.** The interview is one of the most powerful tools for building engagement. Choose a member of the month and interview him or her. Choose your oldest member and interview him or her. And equally important, do VIP interviews. One quick note of advice on VIP interviews: Unless you have an existing relationship, work your way up to the top. Start by finding VIPs slightly lower down on the totem pole—these folks are interviewed far less than CEOs and are often much more excited to talk to the media.

4. **Advice.** This can certainly include advice from the founder, as people do like to hear from the fearless leader, but you can also solicit advice from members and do a roundup of general advice from the community or—very underutilized but useful—advice from those in an adjacent field.

5. **The Guest.** Whether we're talking about op-ed pieces or guest blogs from experts, giving outsiders a forum to communicate with your community can help serve the core and enlarge membership. But it's also worth pointing out that people are busy and many find nothing more intimidating than the blank page— so offer to cowrite pieces as well (though be prepared to do the lion's share of the cowriting).

Engagement and Engagement Strategies

There are two types of engagement that matter most. The first is low-friction engagement, such as a Facebook like or a re-Tweet. The only reason this type of engagement matters is cosmetic. Many newcomers want social proof that the community they've stumbled upon is the real deal, and having 10,000 or more likes on Facebook will help. That said, a like is not deep engagement. As opposed to low-friction engagement, deep engagement requires building living bridges between members of your community. A living bridge means people are connecting with you and connecting with each other, and in ways that generate real emotions. This is critical. People join communities for the ideas; they stay for the emotions.

Now, clearly there are many ways to generate emotions in your community. We'll examine some of the more powerful below, but the most important thing to know is that deep engagement demands rapid experimentation. Remember the point of these experiments is to get people talking to one another and get them working together. Keep trying different ways to make communication easier and increase chances of collaboration.

Here are five of the most useful engagement strategies.

1. *Reputation.* We saw with TopCoder how effectively a rating system and leaderboard drove engagement. Consider that there are
   now dozens and dozens of software companies that won't hire new
   talent unless they have a TopCoder rating. When an engagement
   strategy becomes a business fundamental, that's serious leverage.
   A leaderboard, meanwhile, allows you to add a game-layer to
   the community. Publically holding people accountable for their
   performance creates interesting social dynamics. For competitive members, these dynamics inspire them to work harder to
   improve their spot on the leaderboard. For the less competitive,
   having a leaderboard is a great way to identify areas of expertise

among community members. Remember that it doesn't always have to be complicated. Simply highlighting members' contributions or achievements also enhances their reputation.

2. **The Meet-up.** The goal is to generate real emotions, and nothing works better than live bodies in a room together. Even better, if you can figure out how to make these meet-ups self-organizing—such as Sheryl Sandberg's Lean In circles—you are getting all the benefits of deep engagement with far less effort. Of course, if you can't get everybody together physically, get them together virtually, though don't be afraid of hosting a structured discussion. People are busy. Drawing up boundary lines and focusing the conversation is a great way of showing folks that you respect their time.

3. **The Challenge.** Whether it's an incentive prize (see the next chapter) or a group project or a well-crafted debate, challenging the community can be a great way to foster cohesion. And have challenges within challenges. Use deadlines to keep things interesting. Add rules that require collaboration—for example, a project must be viewed by a specific number of community members before being accepted as a competition entry.

   Equally important, challenges are necessary because they help you keep "entitlement" to a minimum. "The goal of every community is to create a sense of belonging," says Jono Bacon, the Senior Director of Community at XPRIZE. But there's a flip side: the opposite of belonging is 'entitlement.' Many communities struggle with entitlement, and it can cause them to become stale when entitled members slow down the pace of innovation. All communities are at risk of becoming stale when they don't challenge themselves."[24]

4. **Visuals.** Whether it's community founder generated how-to videos or user-generated photos or a simple slideshare, ignoring the fact that the web is a visual medium will only hurt you. People expect a certain degree of eye-candy online today. And eye-candy is easy to crowdsource and easy to share.

5. *Be a Connector.* As the community organizer, you'll likely have access to information about the interests, activities, and backgrounds of your members. One of the best ways to engage them and create immense value in the group, especially in the early days, is to introduce like-minded members to each other. Make the introduction, suggest that they meet, and give them a topic or agenda to fuel the conversation. Then watch the ripple effect spread.

Managing Your Community

Communities are messy places. Yet you need to steer the ship no matter how turbulent the storm. There is a cornucopia of steering wisdom around, but here are the five lessons for managing a community that matter most.

1. *Benign Dictators.* Everyone we talked to said the same thing: The best communities are run by benevolent tyrants. As Local Motors founder Jay Rogers explains, "There are certain times you need to be a benign dictator. For us, we knew we were going to make a car but had to decide which car to make. We could have let the community decide, but it wasn't that clear-cut. We were worried that people would choose a design for intellectual or academic reasons, but the choice would not fit our business model. And we needed to make something that would sell. So we decided to establish parameters and then let our community make suggestions. We also reserved the right to make final decisions. We were transparent about it. We were benign dictators, but we still needed to be dictators."

2. *Stay Calm.* Let the kids play. Will it get loud occasionally? You betcha. But a little fighting is a good thing. So is a little meandering. Writing for Mashable, technology commentator Jolie O'Dell explained it like this: "Often, we jump in too quickly

when a conversation we've started might actually need to simmer for a few hours without our intervention. People need to go off topic, trolls need to be smacked down by power users, sidebar chats need to occur, often without direct comment from within the organization. No one likes the idea that they're being monitored all the time."[25]

3. *No Panhandling.* Stop trying to market things to your community. You're there to support them and not to sell to them. The marketplace emerges organically, from the conversation, and not the other way around.

4. *Retention Matters.* Too many community leaders spend all their time chasing new members. Don't. In DIY communities, bigger is not always better. Plus, if you're constantly trying to increase membership, you're neglecting the members you have—which is an easy way to lose them. Retaining the members you have, making certain they are happily engaged. That's far more important.

5. *Delegate.* Distributed leadership is key. Let community leaders emerge, and be sure to spread power around. Find your best blog poster and put him or her in charge of the comments section. Find a friendly power user and put that person in charge of greeting new members. Delegate contests and research projects and everything else. And do this with authority. You're the benevolent dictator, so establish guidelines and clear responsibilities, provide training when needed, and create perks to reward all this participation.

Driving Growth

Remember, you don't need to be huge to be effective, but if you're looking to grow, the best place to start is with the basics. "People like talking to one another," author Seth Godin (who himself has an enormous community) once said. "We evolved to want to do that. So one of the most highly leveraged and powerful ways to grow a tribe is to connect people to each other. But, if you just have that, you have nothing but a coffee shop. On top of *that*, there needs to be a message from you, the leader, about where you want to go, about the change that you want to make in your world. You need a mission, a movement, a place that people want to get to."[26] In other words, you're not going to grow without a clearly defined MTP and a place for people to get together to try and attain that MTP.

With these basics in place, here are seven effective strategies for expansion.

1. *Evangelism.* Word of mouth is still the most effective way to grow a community. Get your members talking about your efforts. To help create early interest in Local Motors, the staff visited sites frequented by car designers. Jay Rogers explains: "We'd simply say, 'We're going to make a car that you guys design. What do you think?' The important thing is to plant the flag, tell people what you're going to do."

2. *Play Well with Others.* Partner with neighboring organizations. Do this in the real world; do this in cyberspace. One of the reasons TopCoder exploded in membership was because they partnered with Sun Microsystems—with Sun both providing more members and the validation that this community was doing something special.

3. *Competition.* People love to compete. Leaderboards, rating systems, incentive prizes, whatever—give people a way to square off against one another and they will show up.

4. *Pick a Fight.* One of the best ways to strengthen a community is to go into battle against a common rival. Find an enemy. Take a stand.

5. *Buzz Marketing.* Edgy demonstrations of new tech/products/ ideas spark buzz and attract followers. Better Blocks, for example, creates community improvement flash mobs. They band people together to paint bike lanes on streets, plant trees in public spaces, and create outdoor cafés and pop-up shops—all with-

out governmental approval. Not only does this help them build their community, the point they make with these crowdsourced, temporary urban improvements usually leads to changes in legislature and long-term urban renewal.[27]

6. **Host Events.** This has been discussed before, but it's worth repeating: nothing brings people together like, well, actually bringing people together.

7. **Technical Optimization.** If you want a larger online presence, don't forget the tried and true: search engine optimization tactics, AdWords, Facebook advertising, etc.

Monetization

Okay, so you are, after all, an entrepreneur, and making money—at some point—matters to you. Monetizing your community can be more art than science, but there are several hard and fast rules worth remembering.

1. **Transparency and Authenticity.** DIY communities are built on openness, so if you plan on making money from your community, don't hide this fact. Put it in your mission statement. Post it on the site. Everyone we interviewed agreed that being forthright about money means less trouble down the line. Moreover, there's a good chance your community is also looking for ways to make money from their passion, so drive engagement by making monetization a topic for discussion.

2. **Sell What the Community Builds.** The easiest way to make money without alienating members is to help those members make money too. This is the strategy that worked for Local Motors, TopCoder, and a great many others. And if your community isn't building products, they are still building up expertise. You can sell this too via guides, summaries, ebooks, lectures, podcasts, whatever.

3. **Cater to the Core.** Give the people what they want. Sell products that are authentic and do so after you have an established reputation. Chris Anderson waited years before trying to monetize DIY Drones, and when he did, it was by offering to build what his community had designed but (in some cases) had neither the time nor the resources to actually build—fully assembled quadcopters.

4. **All the Typical Stuff.** You can, of course, sell ads to outsiders and premium membership to insiders—these are typical approaches—but again, remember to cater to the core. Make sure your advertisers are selling things the community really wants. Similarly, selling premium memberships can work, but make sure that membership really has privileges and that those privileges don't detract from your established community. Giving paid subscribers access to job boards works great. Giving paid members access to events does as well, but know that if discussion boards end up dominated by insider chatter—that is to say, you had to be there to understand—then people who didn't attend the event aren't going to stick around for long.

Last Words

I want to close out this chapter by mentioning that two of the exponential crowd tools discussed in this section are themselves mechanisms to turn a crowd into a community. The first is crowdfunding. When a crowdfunding campaign is successfully completed, all those who have pledged are now part of your community. At the conclusion of the ARKYD Campaign, we had 17,000 new members to collaborate on with our MTP.

The second mechanism is where we're going next. It's the topic of our final chapter, the incredible innovation accelerator and community-building strategy that helped launch my career: incentive prizes.

CHAPTER TEN

Incentive Competitions

Getting the Best and Brightest to Help Solve Your Challenges

This final chapter focuses on one of the most powerful mechanisms available to the exponential entrepreneur for solving big and bold global challenges, a tool used by powerful companies and successful entrepreneurs. This exponential crowd tool is the *incentive competition*, an idea that combines all of the lessons discussed throughout this book and taps into the most power force in the human psyche, our search for significance.

An incentive competition is straightforward. Set a clear, measurable, and objective goal and offer a large prize to the first person to achieve it. As we shall see, this mechanism pulls together most of the knowledge from the previous nine chapters: the use of exponential technologies, thinking at scale, crowdsourcing genius, providing opportunities for crowdfunding, and stimulating the creation of DIY communities. Moreover, incentive competitions are brutally objective. They don't care where you went to school, how old you are, or what you've ever done before. Billion-dollar corporations compete as equals against two-person start-ups. They measure only one thing: Did you demonstrate the target goal of the competition?

For the exponential entrepreneur, the incentive prize is a mechanism for solving a personal challenge or a global injustice or bringing a new technology into existence. As I mentioned earlier, my original use of incentive competitions stemmed from my desire to figure out how to get myself into space. I had given up on NASA being my ticket, instead turning to commercial space flight as a way to develop both the technology and the wealth needed to get myself off-earth. But there was another impetus as well.

In 1993, I received a copy of Charles Lindbergh's 1954 Pulitzer Prize–winning book, *The Spirit of St. Louis*. This gift came from my dear friend Gregg Maryniak, who was then hoping to provide the inspiration needed for me to finish my pilot's license (which I had started and stopped three times for lack of money and/or time). And it worked. I did complete my license, but the inspiration didn't stop there.

Before I read *Spirit*, I'd always believed that Lindbergh woke up one day and decided to head east, crossing the Atlantic on a whim. I had no idea that he made his famous flight to win the Orteig Prize— a $25,000 prize for the first person to fly solo from Paris to New York (or vice versa). Nor did I know what extraordinary leverage such competitions could provide. In this case, cumulatively, nine teams spent $400,000 trying to win Raymond Orteig's purse. That's sixteenfold leverage. And Orteig didn't pay one cent to the losers: instead his incentive-based mechanism automatically backed Lindbergh, who was, by most accounts, the least qualified of all the entrants. Even better, the resulting media frenzy created so much public excitement that an entire industry was launched. It was an incentive prize that led to today's $300 billion global aviation market.[1]

By the time I finished reading *The Spirit of St. Louis*, the concept of an incentive prize for the "demonstration of a suborbital, private, fully reusable spaceship" had formed in my mind. Not knowing who my "Orteig" would be, I wrote " 'X' PRIZE" in the margin of the book. The letter X was a variable, a placeholder, to be replaced with the name of the person or company that put up the $10 million purse. How I decided on $10 million as the purse size, raised the money, and created the rules, I'll get to shortly. My first step, after realizing that an incentive prize might help me fulfill my personal moonshot, was to learn everything I could about prizes, their history, and how and why they worked.

The Power of Incentive Competitions

Orteig didn't invent incentive prizes. Three centuries before Lindbergh crossed the Atlantic by plane, the British Parliament wanted some help crossing the Atlantic by ship. In 1714, the £20,000 Longitude Prize was offered to the first person to accurately measure longitude at sea. It worked. In 1765, horologist John Harrison pulled it off, but beyond opening the oceans to navigation, this competition brought incentive prizes—as a method for driving innovation—into the public eye.

The idea spread quickly. In 1795, for example, Napoléon I offered a 12,000-franc prize for a method of food preservation to help feed his army on its long march into Russia. The winner, Nicolas Appert, a French candy maker, established the basic method of canning, still in use today.[2] In 1823 the French government offered another prize, this one a 6,000-franc purse for the development of a large-scale commercial hydraulic turbine. The winning design helped power the burgeoning textile industry. Other prizes have driven breakthroughs in transportation, chemistry, and health care.[3] "Historically, for both royalty and industrialists, incentive prizes have long been a tool for fostering innovation," says Deloitte Consulting Principal Marcus Shingles. "But it is only now that these competitions are beginning to reach their prime. In our hyperconnected world, with the maturation of social media and the explosion of crowdsourcing capabilities, our ability to design and utilize these prizes to drive breakthroughs has never been stronger."[4]

The success of these competitions stems from a few underlying principles. First and foremost, large incentive prizes raise the visibility of a particular challenge, attracting innovators and nontraditional thinkers from around the globe. These competitions also help foster the belief that a given challenge is in fact solvable. Considering what we know about cognitive biases, this is no small detail. Before the Ansari XPRIZE, few investors seriously considered the market for commercial human space flight; it was assumed to be the sole province of governments. Afterwards, a half-dozen companies formed, well over $1 billion was invested, and hundreds of millions of dollars' worth of tickets to space have been sold.[5]

Second, in areas where market failures have hindered investment or entrenched incumbents have prevented progress, prizes break bottlenecks. By creating a race with a large cash payout, these competitions attract new forms of money to the problem area. Rather than backing a potential team solely for the investment opportunity, corporate sponsors and benefactors support a team for the publicity. Consider that every year, sponsors spend $45 billion backing teams whose sole purpose is to move different-sized and -shaped balls up and down fields.[6] In a very similar fashion, corporations can now support teams trying to solve grand challenges.

Corporate Sports Sponsorship by Category

* Division 1 Schools Athletic Revenue: Source: Sportsbusinessjournal: Bizofbaseball.com: NCAA Data: Deloitte: Research

Big Business: Corporate Sports Sponsorships

Source: http://www.sportsbusinessdaily.com, *http://www.bizofbaseball.com,* www.deloitte.com

The next factor behind the wild success of incentive competitions is their ability to cast a wide net. Everyone from novices to professionals, from sole proprietors to massive corporations, gets involved. Experts in one field jump to another, bringing with them an influx of nontraditional ideas. Outliers can become central players. At the time of England's Longitude Prize, there was considerable certainty that the purse would go to an astronomer, but the winner, John Harrison, was a self-educated clockmaker.[7] Along similar lines, in the first two months of the Wendy Schmidt Oil Cleanup XCHALLENGE, some 350 potential teams from over twenty nations preregistered for the competition, including one that had come together in a Las Vegas tattoo parlor and had never been involved in the oil cleanup business before (more on this in a moment).

The benefits don't stop here. Because of the competitive framework, everyone's appetite for risk increases, which drives further innovation. Moreover, since many of these competitions require significant capital to field a team, crowdfunding can now be used to attract the requisite financial support—unlocking a potentially global field of backers. Finally, competitions inspire hundreds of different technical approaches, which means that they don't just give birth to a single-point solution, but rather to an entire industry.

Why Prizes Work

The American anthropologist Margaret Mead once said, "Never doubt that a small group of thoughtful, committed citizens can change the world. Indeed, it is the only thing that ever has."[8] As we saw earlier, this same concept was echoed in Kelly Johnson's third rule of skunk: "The number of people having any connection with the project must be restricted in an almost vicious manner." There are pretty good reasons for these opinions. Large or even medium-sized groups—corporations, movements, whatever—aren't built to be nimble, nor are they willing to take large risks. Such organizations are designed to make steady progress and have considerably too much to lose to place the big bets that certain breakthroughs require.

Fortunately, this is not the case with small groups. With no bureaucracy, little to lose, and a passion to prove themselves, when it comes to innovation, small teams consistently outperform larger organizations. And incentive prizes are perfectly designed to harness this energy.

There is another powerful psychological principal at work here: the power of constraints. Creativity, we are often told, is a kind of free-flowing, wide-ranging, "anything goes" kind of thinking. There's an entire literature of "think outside the box" business strategies to go along with these notions, but, if innovation is truly the goal, as brothers Dan and Chip Heath, the bestselling authors of *Made to Stick: Why Some Ideas Survive and Others Die*, pointed out in the pages of *Fast Company*, "Don't think outside the box. Go box shopping. Keep trying on one after another until you find the one that catalyzes your thinking. A good box is like a lane marker on the highway. It's a constraint that liberates."[9]

In a world without constraints, most people take their time on projects and assume far fewer risks, while spending as much money as you'll give them. They try to reach their goals in comfortable and conservative ways—which, of course, leads nowhere new. But this is another reason why incentive prizes are such effective change agents. When you tell someone that they have only a tenth the budget and a tenth the resources (or put conversely, you have to achieve 10x bigger results with the same resources—aka moonshot thinking), most people give up and say it can't be done. A few venturesome entrepreneurs may decide to give it a shot, but if they are paying attention, they'll understand from the outset that the same old way of solving the challenge will no longer work. The only option left to them is to throw out past experiences and preconditions and start with a clean sheet of paper. And this is exactly where serious innovation begins.

Let's take a quick look at how the XPRIZE capitalized on the power of constraints. For starters, the prize money defines spending parameters. The Ansari XPRIZE was $10 million. Most teams, perhaps optimistically (and who would pursue a space prize without being an optimist?), told backers they could win for less. In reality, most teams go over budget, spending considerably more than the prize money trying to solve the problem (because by design, there's a back-end business model in place to help them recoup their investment). But this perceived upper limit tends to keep out risk-averse traditional players. In the case of the XPRIZE, my goal was to dissuade the likes of Boeing, Lockheed Martin, and Airbus from entering the competition. Instead, I wanted a new generation of entrepreneurs reinventing space flight for the masses—which is exactly what happened.

The time limit of a prize competition serves as another liberating constraint. In the pressure cooker of a race, with an ever-looming deadline, teams must quickly come to terms with the fact that the same old way won't work. They're forced to try something new, pick a path, right or wrong, and see what happens. Most teams fail, but with dozens or hundreds competing, does it really matter? If one team succeeds within the constraints, they've created a true breakthrough.

Having a clear, bold target for the competition is the final important restriction. This massively transformative purpose (MTP) galvanizes passion, attracting the best talent and inspiring them to give it their all. In the case of the $30 million Google Lunar XPRIZE, when it was launched in 2007, only two nations had ever landed on the Moon, and no one had been there in more than thirty years. The MTP of the Google Lunar XPRIZE was to enable a new generation of exponential entrepreneurs to build spaceships at one hundred times lower costs to open the space frontier. Over twenty-five teams, comprised of the best and brightest from around the world, entered the competition.[10]

Taken together, as three centuries' worth of history shows, because of the harnessing of passion, the freedom from bureaucracy, and the power of constraints, incentive competitions are one of the most potent innovation turboboosts available.

Case Study 1: Wendy Schmidt Oil Cleanup XCHALLENGE

In April 2010, British Petroleum's Deepwater Horizon oil rig exploded and sank off the Gulf Coast of the United States, causing the largest accidental ocean oil spill in the history of the petroleum industry. Before it was capped, the leaking Macondo Prospect well spewed more than 200 million gallons of oil into the sea, exceeding the infamous 1989 *Exxon Valdez* spill some eighteen times over. The resulting slick covered 2,500 to 4,000 square miles of the Gulf of Mexico, approximately the size of Hawaii's Big Island.[11]

Using a combination of traditional methods, cleanup teams managed to remove less than half the oil, approximately 69 million gallons. Natural dispersal and evaporation removed an additional 84 million gallons. But that left a whopping 53 million gallons, about 26 percent of the spill, to pollute the ocean and adjacent shoreline.

A month later, in May 2010, oil was still gushing into the gulf. The news covered the spill day after day with no end in sight. That was when newly elected XPRIZE trustee, ocean explorer, and Academy Award–winning film producer and director James Cameron emailed me to suggest a rapid-response "flash prize" to address the disaster. Francis Beland, then my vice president of prize development, also an ocean explorer, studied the problem. The idea of a prize to cap the gusher was off the table—BP would never give us (or anyone) access to their data. Next we turned to the idea of impacting the cleanup. We quickly realized that the technology to clean up oil spills had not significantly improved in the twenty-one years since the *Exxon Valdez* spill. In fact, a lot of the equipment being used in the gulf was the very same equipment used decades earlier in Alaska. Why? It turned out to be a multilevel problem, with a number of perverse incentives. Cleanup teams (typically disenfranchised fishermen) often were paid by the hour, giving them no financial reason to be faster or more efficient. Oil companies, meanwhile, had no desire to spend money on better technologies because existing methods fulfilled the minimum requirements set by insurers and regulatory bodies. Finally, there was a lack of pressure from governments and other regulatory bodies to improve oil cleanup technologies. In other words, a long-standing tradition of industry-wide apathy had created the perfect conditions for innovation via competition—a prize to increase the rate of oil spill cleanup on the ocean surface, to try and capture BP's oil before it destroyed the coastline.

I emailed the idea to our entire board of trustees and our biggest benefactors. It read something like: "We are looking to reinvent how we are cleaning up oil spills so the tragedy in the Gulf doesn't happen again. I'm looking for a benefactor to underwrite the purse and the operations for this critical and timely prize." Moments later, I heard from philanthropist Wendy Schmidt, president of the Schmidt Family Foundation and wife of Eric Schmidt, chairman (and CEO at the time) of Google. She offered to underwrite the prize. Less than a day later, we had signed a two-page agreement and were racing toward announcing the $1.4 million Wendy Schmidt Oil Cleanup XCHALLENGE.[12]

To measure success, we decided to lean on two established industry metrics: oil recovery rate (ORR), the amount of oil that can be recovered per minute, and oil recovery efficiency (ORE), the amount of oil recovered per volume of water. For decades, the best ORR had been about 1,100 gallons per minute. To make ours a compelling challenge we wanted teams to—at least—double it. We set the minimum recovery rate at 2,500 gallons per minute, with an ORE of at least 70 percent.

Wendy and I announced Oil Cleanup XCHALLENGE onstage at the National Press Club on July 26, 2010. Very quickly, 350 teams from all over the world preregistered for the competition. Of these, twenty-seven submitted designs by the April 2011 deadline, and ten finalists were selected by our judges, based on five design parameters:

1. Technical approach and commercialization plan
2. No negative impact on the environment
3. Scalability of the technology and ease of deployment

4. Cost and human labor required for implementation
5. Improvement over existing technologies for collecting and
   removing oil

The finalists were an eclectic group. Six teams were comprised of oil industry veterans with established or developing cleanup technologies, while the remaining four were start-ups with little or no oil background whatsoever. Field tests were conducted in the Oil and Hazardous Materials Simulated Environmental Test Tank (OHMSETT) at the National Oil Spill Response & Renewable Energy Test Facility.[13] One of the largest tanks of its kind—measuring 667 feet long by 65 feet wide by 11 feet deep and filled with 2.6 million gallons of salt water—this behemoth simulates real ocean conditions and oil spills in a safe, contained environment, while state-of-the-art data collection and video systems record and test the results. Using OHMSETT, each team made six qualifying test runs, three in calm water and three in wave conditions. The test field was a patch of oil approximately 400 feet long, 60 feet wide, and 1 inch deep, with a volume of 27,000 gallons.

The results were dazzling. Seven teams doubled the industry's previous best ORE. One of the teams, Elastec/American Marine, won first prize by achieving an ORE of 89.5 percent and an ORR of 4,670 gallons per minute, or a 400 percent improvement over the industry's all-time best. (Since the end of the competition, Team Elastec/American Marine has actually improved their ORR, exceeding 6,000 gallons per minute.)

But the most memorable outcome came from one of the finalists who didn't win. Vor-Tek was one of the teams that doubled the oil spill cleanup rate, but didn't place in the top three. They were a team of complete novices from far outside the oil cleanup business. They had met at a Las Vegas tattoo parlor. The technology designer was a tattoo artist, his customer funded the work, and to test out their ideas, they built a scale model in a Jacuzzi. The first time their technology saw full-scale oil and water was at OHMSETT, and they still doubled the preexisting cleanup rate. When asked about their experience, Vor-Tek member and tattoo artist Fred Giovannitti said, "We get asked all the time, 'How long have you been in the oil industry?' and I ask back, 'Counting today?'"

The lesson here is that in incentive competitions, results can come from the most unusual of places, from players you would never expect, and from technologies you might never suspect. Lee Stein, an XPRIZE benefactor, says, "When you are looking for a needle in the haystack, incentive competitions help the needle come to you."

Case Study 2: The Netflix Prize

The best incentive prizes are those that solve important puzzles that people want solved and people want to solve—and there's a difference. The Wendy Schmidt Oil Cleanup XCHALLENGE falls directly into the former category. It took me over ten years to raise the money for the Ansari XPRIZE, but Wendy Schmidt stepped forward to fund the Oil Cleanup Challenge in less than forty-eight hours. Certainly one reason I raised money for the Oil Cleanup Challenge so quickly was the fact that by then I had a track record of success and a considerably thicker Rolodex, but a more important factor was the 800,000 gallons of crude gushing into the Gulf Coast each day. Disaster is a motivator because empathy is a motivator, and empathy is never higher than when the same disaster movie has been playing on TV for over a month. But my point here isn't about capitalizing on misfortune, it's about capturing momentum.

Every good prize needs this kind of momentum. The Qualcomm Tricorder XPRIZE—a $10 million prize for the first team that can build a handheld device that diagnoses illness better than a team of board-certified doctors—saw 330 preregistered teams from thirty-three countries enter the competition in its first twelve months.[14] Why? Because faster and more accurate diagnosis is a bold common good. It's billions and billions of dollars in health care savings and, in parts of the world where there aren't enough doctors, a matter of life and death.

This means you don't have to capitalize on misfortune to achieve this kind of momentum; capitalizing on a bold vision works just the same.

But when designing a prize, there is a second kind of momentum you can tap into—our innate desire to compete. Take coders. Consider what we learned from meeting Jack Hughes and exploring TopCoder. First, coders are a competitive bunch. They like besting one another and they like leaderboards for bragging rights. But what else do coders enjoy doing? They watch movies—a lot of movies. They're the same bunch that lines up three days early for the latest Star Wars release and stays up three days straight arguing Freddy versus Jason. Thus, if you could create an incentive prize that harnessed this competitive love of coding and this argumentative love of movies and tied them together—meaning design a prize around the intrinsic motivations at the core of coder culture—what might be possible?

Well, in the case of Netflix, a better movie recommendation engine.

A movie recommendation engine is a bit of software that tells you what movie you might want to watch next based on movies you've already watched and rated (on a scale of one to five stars). Netflix's original recommendation engine, Cinematch, was created back in 2000 and quickly proved to be a wild success. Within a few years, nearly two-thirds of their rental business was being driven by their recommendation engine. Thus the obvious corollary: the better their recommendation engine, the better their business. And that was the problem.

By the middle 2000s, Netflix engineers had plucked all the low-hanging fruit and the rate of Cinematch optimization had slowed to a crawl. Every time one of their recommendations was a clear miss—based on your interest in *Breakfast at Tiffany's* we think you'll enjoy *Naked Lunch*—customers got angry. And with new competitors sprouting up in the likes of Hulu, Amazon, and YouTube, this ire was getting expensive. So Netflix decided to attack the problem head-on, announcing the Netflix Prize in October 2006—a million-dollar purse for whoever could write an algorithm that improved their existing system by 10 percent.[15]

And this contest is a perfect example of what happens when you design prizes around intrinsic motivations. Competition, coding, and movies—what could be more fun than that? Within two weeks, Netflix had received nearly 170 submissions, three of them outperforming Cinematch. Within ten months, there were over 20,000 teams from 150 different countries involved. By the time the contest was won, in 2009, that figure had doubled to 40,000 teams.

But the results that Netflix saw extended far beyond the number of contestants entered in a contest. As Jordan Ellenberg explained in *Wired*: "Secrecy hasn't been a big part of the Netflix competition. The prize hunters, even the leaders, are startlingly open about the methods they're using, acting more like academics huddled over a knotty problem than entrepreneurs jostling for a $1 million payday. In December 2006, a competitor called 'simonfunk' posted a complete description of his algorithm—which at the time was tied for third place—giving everyone else the opportunity to piggyback on his progress. 'We had no idea the extent to which people would collaborate with each other,' says Jim Bennett, vice president for recommendation systems at Netflix."[16]

And this isn't an aberration. Over the course of the eight XPRIZEs launched to date, there has been an extraordinary amount of cooperation. We've seen teams providing unsolicited advice, teams merging, teams acquiring and sharing technology and experts. When the prize is driven by an MTP, while a team's primary purpose is to win, a close second is their desire to see the primary objective achieved; thus teams exhibit a much higher willingness to share.

A well-designed incentive competition provides teams with a "whole is bigger than the sum of the parts" mindset. This happens because the right motivations lead to increased cooperation, which leads to unpredictable network effects. In November 2007, for example, progress toward Netflix's desired 10 percent improvement had slowed considerably. The geeks had taken things about as far as possible when a British psychologist named Gavin Potter entered the fray. Instead of the purely mathematical approach utilized by most of the other teams, Potter was making startling progress by considering the human factor (he had actually outsourced the difficult math to his daughter, then a high school student). Potter didn't end up winning the competition, but the winning team did begin taking the human factor into account and this helped them to victory.[17]

In the coming years, competitions such as the Netflix Prize will become more and more important. Today's world is awash in data, and mining this treasure trove for useful tidbits can be worth billions of dollars. Tomorrow's world will be even more information packed. As we are entering an era of a trillion sensors and ubiquitous networks, we are going to be able to gather data about anything, anywhere, any time we want. Incentive prizes provide exponential entrepreneurs with an incredibly efficient method to extract tremendous knowledge out of this bounty, providing an innovation acceleration engine unrivaled in history.

Case Study 3: HeroX

I spend a lot of time giving presentations to corporations. When I speak to executives, there are six key points I stress.

1. The only constant is change.
2. The rate of change is increasing.
3. If you don't disrupt yourself, someone else will.
4. Competition and disruption are no longer coming from some multinational company overseas. They now originate from the guy or gal in a start-up garage harnessing exponential technologies.
5. Given Bill Joy's famous comment "No matter who you are, most of the smartest people work for someone else," how do you tap into these individuals?
6. If you're dependent upon innovation only from within your company, you are dead. You must harness the crowd to remain competitive.

While the XPRIZE Foundation has been incredibly successful, a few years back I heeded my own advice and asked myself how I would disrupt my own company. Or more specifically, how might someone disrupt XPRIZE? The answer, which should now be evident, is the creation of an online platform that allows anyone to launch a challenge in any area he or she cares about, where the crowd can help design the prize, fund the prize, and ultimately compete to win the purse. A platform able to disrupt archaic closed innovation systems, where hundreds or thousands of smaller challenges can be launched per year, thus scaling way beyond the three multimillion-dollar XPRIZEs launched each year. Sort of a Craigslist meets Indiegogo scenario.

The platform, now named HeroX,[18] was incubated at XPRIZE headquarters, but, as with any skunk works, separation from the mother ship was critical. So HeroX hired a small, passionate virtual team, with members located everywhere from Canada to Ukraine.[19] The CEO, Christian Cotichini, turned out to be the venture's first major outside equity investor.

As with any new start-up, launching early is critical, so we turned to our existing supporter base for help. Graham Weston, cofounder of Rackspace, stepped up to be our first customer. Weston wanted to help Mexican entrepreneurs open up stores on the US side of the border, specifically in his hometown of San Antonio. To do this, he launched a 24-month, $500,000 competition on the HeroX platform dubbed the San Antonio Mx Challenge.

"The top two obstacles faced by Mexican entrepreneurs are access to visas and access to information," says Weston. "Getting a visa to work in the United States is very difficult. The laws are not designed for tech entrepreneurs. Second, many Mexican entrepreneurs struggle with the tactical requirements of launching a business in the US— around things like financing, recruiting, real estate, and employment law—and do not know where to go to for information and help."[20]

To counter these obstacles, the San Antonio Mx Challenge will pay $500,000 to the individual, team, or organization that creates and implements a repeatable model to assist Mexican tech companies in opening active offices in San Antonio. The winner will score the highest on a points-based system that measures three things: the number of Mexican companies attracted over two years, the total combined revenue of those companies for two years, and finally, the sustainability of those companies and their business models.[21]

On the heels of this successful launch, HeroX is now developing dozens of challenges in dozens of cities. In conjunction with the city of Los Angeles, there's a challenge being designed to decrease traffic on Interstate 405. Simply Music is using HeroX to create a virtual piano so that musical expression is no longer tied to physical instruments. The education software giant Ellucian is using HeroX to increase student retention and graduation rates, and a fifteen-year-old high school student, Eli Wachs, is using HeroX to show the world that young people are change makers.

But the real point here is accessibility. HeroX is helping build a passionate and knowledgeable community of prize developers who can help any entrepreneur design a prize, launch a prize, use crowdfunding to supplement the purse, operate the prize, and ultimately judge and award the prize. The big goal is to change people's mindset—to help them realize that they no longer need complain about problems, but now can launch an incentive competition to solve them.

The Benefits of Using an Incentive Competition

My goals in this section are twofold: first, to help you to identify a prizable topic useful to you and your business, and second, to help you design your own incentive challenge with HeroX. But we'll begin by considering why you might use an incentive prize in the first place. What are the benefits to you, your company, and society?[22]

1. *Attracting new capital to innovators solving the problem.*
   We normally think of government agencies or big corporations as the primary funding sources for innovation. Yet, incentive prizes attract a very different, nontraditional pool of resources to

the innovation game, specifically billions in resources normally allocated for both philanthropy and sponsorship.

2. *You pay only the winner.* Prizes are efficient. They generate an enormous amount of innovation—often enough to create an entire industry—but you have to pay only the winner, none of the teams who attempt and fail. In the case of the Orteig Prize, most of the famous aeronauts of the time failed miserably, while Lindbergh, a relatively unknown pilot who was called the "Flying Fool" by the press, won the competition. Had Orteig been investing in teams, Lindbergh would have been the least likely to get an endorsement.

3. *Crowdsourcing genius.* Prizes attract new players—outsiders, mavericks, and other innovators unlikely to work within a traditional research setting. A properly structured incentive prize will draw on a much larger global talent pool than traditional research efforts, pushing the world's best and brightest minds (independent of age, race, and gender) to work harder, faster, and sometimes collaboratively (on the same team).

4. *Increasing public awareness and raising the visibility of a problem.* The publicity generated by an incentive prize serves an educational function, focusing attention on the importance of the problem. In turn, this global media attention motivates the competing teams to work harder and, in many cases, take bigger risks.

5. *Overcoming existing constraints.* Incentive competitions reconfigure what is possible by transcending societal constraints, legal/regulatory hurdles, and policy regimes. Prizes don't care how old you are or where you work; they measure only the quality of your idea and its execution. As such, solutions constrained by stodgy CEOs or self-reinforcing labor unions can be accelerated into action.

6. *Changing the paradigm.* Incentive competitions help change the paradigm of what people believe possible. Before Lindbergh's flight, aircraft were for aeronauts and daredevils. Afterwards,

they were for passengers and pilots. The general view of transatlantic flight was transformed, paving the way for the emergence
of the airline industry. Before the Ansari XPRIZE existed, space
flight was a game played by governments; afterwards it was open
to anyone.

7. *Launching an industry, with lasting benefit and impact.* An
incentive competition should be designed so that the awarding
of the purse is not the end of the story, but rather the beginning
of a new industry. To this end, innovation alone is not sufficient.
To drive the kinds of breakthroughs that benefit humanity, these
innovations need to be brought to market. Ultimately, the goal
is to solve the problem and stimulate entrepreneurship, bringing
about a new set of products and services that serve as the backbone for a new industry.

8. *Providing financial leverage.* A well-constructed challenge
can easily generate investments an order of magnitude greater
than the purse. Innovators and investors are typically willing to
invest more than the amount of the purse for two reasons. First,
most competing teams are typically optimists. They initially
believe they can win the competition by spending less than the
purse amount and then incrementally rationalize larger investments over time. Second, a properly designed prize has a back-
end business model that allows teams to capture a return on
their investment.

9. *Creation of market demand.* Before the Orteig Prize, there
was no public demand for transatlantic airline flights because
few believed such a crossing was even possible. Such was the
situation with space flight and the Ansari XPRIZE. Successfully
designed and executed incentive challenges create significant
market demand, which tends to establish markets and attract
investment capital.

10. *Attracting new expertise and cross-disciplinary solutions.*
True breakthroughs often come from outside the normal field of
experts. Strongly designed challenges lift a problem to high vis-

ibility, attracting nontraditional innovators and driving interdisciplinary collaboration among unlikely partners.

11. *Driving regulatory reform.* On some occasions, a powerful incentive prize can also drive governmental change, helping to clarify regulatory issues relevant to the competition. The publicity surrounding a prize coupled with a large number of entrants can provide the political pressure needed to foster change. In the case of the Ansari XPRIZE, the competition drove the US Federal Aviation Administration (FAA) to adopt regulations that permitted human space flight in commercial reusable space vehicles.

12. *Inspiration, hope, and intelligent risk taking.* Ultimately, challenges are about fostering innovation and creating hope in fields that are stuck in ruts. If nontraditional teams take intelligent risks in fields dominated by risk-averse incumbents, true breakthroughs are far more likely. Remember, the day before something is truly a breakthrough, it's a crazy idea.

Where Do Prizes Make Sense?

Prizes are not panaceas. Many challenges are too complicated to be prizable and others require teams to raise too much money to compete. In my experience, prizes make the most sense in the following circumstances:

1. *You have a clear understanding of your target, but not the method to get there.* In the case of the Ansari XPRIZE, I knew I wanted a spaceship that could get consumers repeatedly a hundred kilometers into space. I didn't know (or care) what type of propulsion system, landing system, or materials the vehicle would use.

2. *You have a large enough crowd of innovators to tap into.* You want innovators from everywhere. Restricting entrance into

a competition to smaller talent pools produces lesser results. The Wendy Schmidt Oil Cleanup XCHALLENGE attracted 350 teams from around the globe. Had we restricted the challenge only to students at a single university, we would never have achieved our desired goals.

3. *A small team is capable of solving the challenge.* The ideal competition can be solved by a reasonably small team. In the case of the DARPA Grand Challenge for autonomous cars, it was a team of graduate students from Stanford. In the case of the Ansari XPRIZE, it was a group of thirty engineers from Scaled Composites. Projects requiring a team much larger will likely run into fund-raising and management challenges.

4. *You are flexible on timeline, types of solutions, and who might win.* When using an incentive prize solution, you give up a certain amount of control in exchange for getting unexpected, potentially breakthrough results from nontraditional players. If you specify challenge parameters too narrowly—such as which technologies must be used or where the innovators should come from—you lower your chances of getting the results you seek.

5. *You are flexible on who owns the intellectual property at the end.* We'll discuss intellectual property (IP) in greater length below, but in the case of most XPRIZEs, the IP is retained by the winning team, and the prize sponsor backing the competition is doing so for the purpose of publicity or to bring real change to the world. This is not necessarily the case for HeroX challenges, where the IP can be owned at the end by the challenge sponsor.

The Big Three Motivators

As I've studied prizes, I've identified three principal motivators that attract teams to compete. Understanding these principles can help you fine-tune your competition to maximize participation.

1. *Significance/recognition.* There's a lot of latent talent that wants the chance to prove itself to the world. Prizes, especially those high in MTP and visibility, offer the winning team the chance for rapid fame.

2. *Money.* While many teams don't compete only for the money, sometimes the cash can be a real motivator. Such was the case for Dr. Paul MacCready, who designed and built the Gossamer Condor, a human-powered vehicle that flew a figure eight between two markers half a mile apart. MacCready pursued the challenge to win the £50,000 Kremer prize and pay off a personal debt.[23]

3. *Frustration.* In many cases, such as with the Wendy Schmidt Oil Cleanup competition, the competing teams are deeply frustrated by the status quo and want to solve the problem. Thus competition gives them a target to shoot toward, and a way to focus their frustration.

Key Parameters for Designing Your Incentive Challenge

As you design your own incentive competition or HeroX challenge, there are fifteen important parameters to consider.

1. *Simple, measurable, and objective rules.* When creating a challenge, strive for rules that are straightforward, measurable, and objective, with a finish line that makes the winning of the prize obvious to everyone. In the case of the Orteig Prize, the rules were "fly nonstop between New York and Paris." In the case of the Ansari XPRIZE, the simple version of the rules could be expressed as "Fly the same three-person spaceship to a hundred kilometers in altitude, twice in two weeks." Of course the detailed rules are far more complex, but good prizes are easy to explain and understand.

2. ***Define the problem, not the solution.*** The prize rules should define a problem to be solved, not a solution to be implemented. For example, the Ansari XPRIZE did not care about launch-vehicle details (propulsion, landing mechanisms, etc.). The only objective was to get three people to 100 kilometers twice in two weeks. As a result, the competition saw over a dozen uniquely different approaches.

3. ***Pick the appropriate structure.*** Incentive competitions come in a variety of different structures. Here's a list of a few variants worth considering. Find one that is best for you:
   - *Past the post.* This type of competition offers cash to the first team to achieve the set goal.
   - *Past the post with a deadline.* This is how we structured the Ansari XPRIZE. We offered up $10 million for the first person to fly twice to a hundred kilometers altitude *before* December 31, 2004.
   - *Bake-off.* This is most similar to the Olympic Games. A bake-off competition takes place on a certain date, where teams compete head-to-head, and the best performance in the competition is awarded the purse.
   - *Bake-off with a minimum performance threshold.* This is how we structured the Wendy Schmidt Oil Cleanup XCHALLENGE. Teams delivered their hardware to the same location and competed head-to-head. The best performing team, above a minimum performance (2,500 gallons per minute of oil cleaned up), won the competition.

4. ***Addressing market failures.*** Incentive competitions are often needed to jump-start a stuck industry and demonstrate a new market. Prizes should address problems where a market failure prevents solutions. Here are a few examples of common types of market failures:
   - People believe a problem is not solvable. There is institutional and public misperception.

- There is a stigma that prevents people from even attempting to solve the problem.
- Entrenched players or unions prevent fair competition or transformation of the industry or technology.
- Capital is not flowing into an important problem area.
- Regulatory structures prevent the innovation from materializing.

5. *The proper balance of audacity and achievability.* The prize needs to be audacious enough that it is inspirational (i.e., has an MTP), but not so difficult that it can't be achieved. When I originally announced that the goal of the Ansari XPRIZE was a suborbital flight to a hundred kilometers, many criticized the competition, arguing the target should be private flight into Earth orbit. Had the latter been the objective, it's unlikely that the competition would have been won (energy-wise, orbit is fifty times harder than a suborbital hop to a hundred kilometers). In other words, suborbital flight was sufficiently audacious and achievable—we didn't need to go further to change the paradigm.

6. *Purse size.* Purses come in all sizes. A typical XPRIZE purse runs from $2 million to $30 million, while the average HeroX challenge ranges from $10,000 to $1 million. The size of the purse depends on a number of variables: an understanding of the incentive needed to encourage action, the value of the back-end marketplace, the minimum amount needed to attempt the feat (i.e., a purse might be sized according to the minimum expected that a team might spend), the perceived importance of the problem, and the sponsors' desire for branding (the "biggest ever"). Teams are typically willing to invest more than the amount of the purse if the competition has a back-end business model that allows them to recoup their investment. In the case of higher-end prizes, the large prize purse (for example, $10 million) is used to break through the media clutter, raise the visibility of a problem, and attract nontraditional players.

7. *Persistent media exposure over time for prize competition.* The best-designed challenges have a competition structure that produces ongoing media. This consistent attention attracts funders, builds community, and helps produce the desired change in mindset. In the case of the Ansari XPRIZE, the competition required a pair of flights over two weeks. Teams got far more exposure than would have been the case if the competition simply required one flight on a single day. The best prize designs keep the conversation alive from start to finish.

8. *A telegenic and captivating finish.* Competitions with telegenic finishes—that is, a finish that is extremely visually compelling—will help drive media attention, which, in turn, drives teams to spend more time and money in their attempts to win (everyone wants to be famous). Such a finish also drives media impressions, which educate the public about the change created.

9. *Multiple purses and bonuses.* Using multiple purses (e.g., second and third place) and "bonus" purses can increase the number of teams competing and the variety of approaches pursued. Secondary purses can keep teams engaged even if there is a strong front runner, and keep teams competing after first place has been awarded. It can also lengthen the time of the competition, thereby increasing its ability to achieve paradigm change.

10. *Launching above the line of super-credibility.* The initial announcement of your challenge should be highly visible and super-credible. The launch should drive maximum media exposure, both publicizing the prize and its sponsors and ensuring that the competition is taken seriously from the start. Properly done, a super-credible launch changes the public's perception of the challenge from "Can it be done?" to "When will it happen and who will win?" At the launch event, it is important to have the participation of gold-plated endorsers (who share their reputational equity) and a number of teams ready to compete.

11. ***Global participation/open to all.*** The best incentive competitions are global in nature. In seeking the broadest range of qualified teams—independent of age, education, and experience—you maximize the opportunity for breakthrough results. In other words, don't try to anticipate where solutions will come from. In the case of the Longitude Prize, the British Admiralty was so certain that determining longitude would come from looking at the stars, they filled the committee charged with picking the winner with astronomers. As a result, John Harrison, a watchmaker, was denied the purse for nearly a decade.

12. ***Prize timelines and deadlines.*** The prize timeline is a function of the competition's degree of difficulty. Smaller HeroX challenges might be awarded in six months to a year, while larger-end $10 million XPRIZEs are designed to be won in a three- to eight-year time frame. Couple an appropriate timeline with a deadline, and you'll get increased action. The Ansari XPRIZE, launched in May of 1996, took eight years, and was won less than three months before the December 31, 2004, deadline.

13. ***Ownership of intellectual property (IP) and media rights.*** In a typical XPRIZE, the teams retain IP and the XPF retains media rights. Other prize designs may require that the IP be made available to the public or that a portion of the IP is owned by or licensed to the prize sponsor. If the prize sponsor gets the IP in exchange for the prize purse, then it's a commercial prize. If the IP is retained by the team or put into open domain, then the prize is considered philanthropic and the prize purse is typically tax deductible. In general, there are four variants worth considering:
   - *Philanthropic.* Winner retains IP.
   - *Philanthropic.* IP is put into open domain.
   - *Commercial.* IP is owned by the prize sponsor.
   - *Commercial.* IP is licensed (or shared) by the prize sponsor.

14. *Incorporation of a back-end business model into the prize design.* The ideal competition is designed so that there is a back-end business opportunity for teams to exploit once the prize is won. For example, the Ansari XPRIZE required a three-person spaceship rather than a one-person ship. This opened the possibility of space tourism, allowing for a commercial business model that made it easier for teams to raise funding and was one of the main reasons they were willing to spend far more money than the purse in their attempts to win. When a team wins, the resulting publicity drives capital investment, deployment of the technology, market acceptance, and a new industry that produces a long-term solution to the market failure initially targeted by the competition.

15. *Writing the final set of rules.* The rules are an incentive competition's DNA—they will determine the competition's success or failure and its validity over time. Rules that are made invalid by changing technology or political/social conditions are problematic. Rules that are naive or easily broken can lead to negative or empty results. Consider the case of Nobel laureate Richard Feynman's famous prizes announced during his 1959 lecture "There's Plenty of Room at the Bottom." Feynman offered two $1,000 prizes, one for the first person to build a working motor within a one-millimeter cube and the second for the first person to write the information from a book page on a surface 1/25,000 smaller in linear scale.[24] The rules for his first prize were poorly conceived. While Feynman was looking to promote nanotechnology, what he received instead was a working motor built by an enterprising graduate student using meticulous craftsman skills and conventional tools (jeweler's tweezers and a microscope). Feynman paid the prize but didn't achieve his goal. That said, in 1985, Tom Newman, a Stanford graduate student, successfully captured the second Feynman prize by reducing the first paragraph of *A Tale of Two Cities* by 1/25,000.[25]

• • •

Today, the XPRIZE spends a lot of time thinking about fundamental objectives. How to achieve a goal without specifying the exact process and how to avoid a false wins (i.e., that micro-motor built from conventional tools). At the beginning of a competition, we propose a set of guidelines that are publically distributed and open for comment. There is extensive discussion with the teams and then, months later, the guidelines are converted into a final set of rules. At the unveiling of SpaceShipOne, Burt Rutan noted, "It's amazing that the rules for the XPRIZE are still valid today, nearly eight years after they were announced in 1996." This was an important lesson.

The Step-by-Step How-To of Your Incentive Competition

With all of the parameters above in mind, now it's time to design, build, and launch your own challenge. The HeroX Platform can help you do this, or you can make it happen on your own. Here are the steps involved:

1. IDEATION. WHAT IS THE PROBLEM YOU WANT TO SOLVE?

Identify the key issue. What problem keeps you up at night? It can be technological, social, or market-based. What paradigm do you want to change via competition? What might the world look like after the prize is won? Work to identify the market failures that led to this impasse. Peel back the layers and determine what's at the core. This ideation phase will help you to identify which parts of the problem are best to focus on solving.

2. GUIDELINES AND METRICS. WHAT PARAMETERS ARE YOU MEASURING?

Your next step is to define the key attributes of success. What do you want teams to achieve during the competition? What does the finish line look like? What are you measuring? How will you measure it? Is it cost-free to judge your competition or labor-intensive? Is the target too audacious? How will the public (or your community) perceive the objective you set?

3. THE OTHER DETAILS: NAME, PURSE, DURATION, AND IP

- *Name.* What will you name your challenge? You're looking for something recognizable, easy to remember, cool, and snappy—something that captures the essence of the competition and spreads virally.
- *Prize purse.* How big will you make your purse? How much is the solution to this problem worth? The goal is a purse large enough to attract innovators, but not so large that it encourages the old guard to compete. The right purse is typically enough to cover the baseline costs an innovative team might spend. Also, if you're short on cash, crowdfunding is an option, but be sure to choose a prize name, objective, and MTP that motivates your community to contribute.
- *Duration and format.* Most prizes should have a deadline (which you always have the option to extend). How long do you think it should take to solve the problem? Remember, shorter deadlines drive greater risk taking, but deadlines that are too short can keep teams from entering. What is the best structure for your challenge? Will you award the first team to achieve a minimum threshold? Will you make it an annual competition, in which you reward the best performance each year on a certain

date (think Olympics)? Each approach can bring great benefits, but each has different cost implications for operating the challenge.

- *Intellectual property.* Who owns the IP at the end of your competition? If you don't have to own it and can allow teams to retain the innovation, then you may get more teams competing. You can alternatively ask them to give you a license or put it into the public domain.

4. POLISHING YOUR PRIZE DESIGN

Before you launch your prize, take the time for one more rules review. Here you want to optimize the competition around the following parameters:

- Make it hard to cheat. Remember the Feynman prize example, where a graduate student built a micro-motor with tweezers. Can you improve your rules to prevent this sort of cheating or false win?
- Examine the rules to make sure that the key indicators are sufficiently objective and measurable. Put differently, make sure you know how to pick a winner. Can your judges easily determine success, or will it require expensive and exotic equipment? Answering these questions ahead of time will save you considerable heartbreak later.
- Have you estimated how much it will cost to run the competition? Can you figure out cheaper ways to host it, judge it, promote it?
- When you explain your competition to your friends, do they clearly understand what a team needs to do in order to win? Can a child explain it to his/her parents across the dinner table? Do you have an easy-to-communicate one-line pitch?

- Is the winning moment of the competition sufficiently telegenic that big media will be interested? Or have you designed a boring competition where winning of the challenge is determined by a single bit changing from a zero to a one at the end of a printout?
- If your prize is actually won, will the winning technology actually cause the impact you desire? Will it solve the preexisting market failures? Will it birth a new industry?

5. LAUNCHING YOUR CHALLENGE, REGISTERING TEAMS

Your next goal is to launch your competition above the line of super-credibility. You need to get media and social networks buzzing about the challenge, and you need to create an easy mechanism for newly excited teams to register for the competition. You also need to consider where teams might come from—universities, small companies, your employees, your local community—and make sure your launch is aimed at the right communities. Remember, teams are the core of your incentive competition; recruiting them and meeting their needs are paramount to success.

6. OPERATING YOUR CHALLENGE

Most people don't realize that operating an incentive challenge is not free. In fact the cost of operating an XPRIZE is often equal to the purse itself. Running the competition, interfacing with the teams, dealing with the legal work, making sure the playing field remains level, handling the PR, and so forth requires staff and time. The process can stretch from months to years. Make sure you have the resources in place to meet such challenges.

Even with HeroX, which offers a platform that facilitates most of these requirements while significantly reducing operating costs, challenges still require the following elements to succeed:

- *Legal.* To register, teams must sign a robust agreement that outlines the rules of the competition and what happens under different circumstances. In my experience, the best practice is to create an easy mechanism for newly excited teams to express interest in the competition. A simple form that gathers their contact details works well. Participants then need to sign on to a Master Team Agreement, prepared by your legal counsel, which is effectively a contract that stipulates what a team needs to do to win, what rights you retain and what rights they retain.
- *Prize Lead.* Someone to be the face of the competition, who can speak about the vision and the mission and field the hard questions that always emerge.
- *Community and Team Manager.* The person who engages with the teams and the community as a whole. They're there to answer all questions and ensure the competition produces maximum impact.
- *Judges.* A group of individuals who are completely independent who will help you determine a winner.

7. JUDGING, AWARDING, AND PUBLICIZING

The final phase of your competition involves determining the winner. Judging involves making sure that you, all the teams, the media, and the public know—in a noncontroversial manner—who won and why.

Next is the awarding of the purse (and trophy, etc.). Here the goal is to maximize the promotion of the winning moment. Celebrate the winner(s), sponsors, judges—really, everyone involved. The goal is to effect deep change. This can be accomplished only via exposure. A lot of people need to know this seemingly impossible challenge is now solvable. This is why having a telegenic finish is so important.

Closing Thoughts

Over the course of the last few years I've defined my own massively transformative purpose. After a few iterations and false starts, I find that I'm happiest with the following: "To help entrepreneurs create extraordinary wealth while creating a world of abundance." This MTP comes from the realization that the world's biggest problems are also the world's biggest business opportunities. These problems are modern-day gold mines. The bigger the issue, the more valuable and important the solution.

And the number of players in the world who are able to mine this gold and take on such challenges has exploded. A few hundred years ago, such activities were solely the domain of royalty. A few decades ago, they belonged to national leaders and the heads of multinational corporations. But today, almost anyone with a passion has the power to bring real change into this world.

Ultimately, that has been the point of this book. The exponential technologies discussed in part one give us the physical tools for radical change, the psychological strategies described in part two are the mental framework for success, and the exponential crowd tools that fill part three provide all of the additional resources (talent, money, and so forth) needed to cross the finish line.

Here's the most important point: Abundance is not a techno-utopian vision. Technology alone will not bring us this better world. It is up to you and me. To bring on this better world is going to require what could easily be the largest cooperative effort in history. In other words, there is a bold and bright future out there. But, as with everything else, what happens next is up to us.

And this brings me to my final thoughts. In *Abundance*, Steven and I closed the book with a section on the dangers of exponentials. This time we're turning our attention to leadership. The importance of this topic was raised by Marcus Shingles, an innovation leader from Deloitte Consulting whom we met earlier in this book. "The coming age of exponentials will put game-changing technologies in the hands of everyone," said Shingles. "And while this will no doubt lead us down the road to abundance, it also has the potential to concentrate wealth and power in the hands of the few. To navigate the turbulent times ahead, we will need a new breed of ethical leaders who are not corrupted by such absolute power."[26]

Shingles's call for a new caliber of moral leadership is indisputable and well timed. And while this book has been about bold entrepreneurship and bold impact, we want to close with a call for bold leadership.

Most evil is done in the dark. Dictators and despots oppress women, children, and minorities in secret, when few are watching. But, in the exponential times ahead, in a world of a trillion sensors, drones, satellites, and glass, someone will always be watching. While this raises serious concerns for privacy, it also offers us hope for the end of oppression and perhaps the beginning of an entirely new breed of moral global leadership.

Who will be the Martin Luther King, Jr., or Mahatma Gandhi of the exponential age? Our history tells us that this breed of leaders is extremely rare and often underappreciated at first glance. Perhaps such leadership will materialize from experimentation in virtual worlds, or emerge from some crowdsourced competition, or be yielded over to a benevolent artificial intelligence. Each is, for the first time ever, a real possibility. Perhaps such leadership will arise the old-fashioned way, from those few concerned citizens willing to suffer the long and lonely hours it takes to see farther and hope further and build bridges across the seemingly vast chasms that still so frequently divide us.

One thing is for sure, in those immortal words of Voltaire (famously stolen by Stan Lee for *Spider-Man*): "With great power comes great responsibility." And each of us, like it or not, are now the recipients of great power. This means we now have the power to solve the world's grand challenges and create a world of abundance. But this also means triumphing over age-old bad habits. Greed, fear, slavery, cruelty, tyranny—haven't these curses outlived their usefulness?

Think about how far we've come. Shelter is among our oldest needs, yet right now we can 3-D print ten single-family homes in a day. Health care ranks right beside shelter. And sometime in the next five years, we'll be able to diagnose disease via AI—and thus democratize health care. We have been reaching for the stars for as long as we have been able to tilt our heads upward and gaze in wonder. And sometime in the next ten years we are going to launch our first asteroid mining mission. No doubt about it, we are a species built for bold. But without bold leadership to help us set the course, our history also tells us that we can wander in the desert of bad decisions for a mighty long time.

"The adjacent possible" is theoretical biologist Stuart Kauffman's wonderful term for all the myriad paths unlocked by every novel discovery, the multitude of universes hidden inside something as simple as an idea.[27] Abundance is one of those simple ideas. Its time has come. It is up to the bold to unlock this adjacent possible, to help humanity live up to our full exponential potential.

AFTERWORD

NEXT STEPS—HOW TO TAKE ACTION

It's an exciting time. Every week, there are new technologies slipping out of the lab and into the market, driving us toward a world of abundance. We think it's critically important for you to have access to this ever-expanding pool of knowledge, therefore we're presenting five different ways for you to stay plugged in, interact with the authors, and join an ongoing conservation about the radical advances bringing us a world of abundance.

AbundanceHub.com: Free, Up-to-Date Content

Visit our website, www.AbundanceHub.com, where you'll have access to up-to-date data, articles, blogs, and videos about abundance and exponential technologies. This site is free and media rich. You also can sign up for a free newsletter and participate in future initiatives, get weekly blog posts on "New Evidence of Abundance," and much more.

Personal Coaching from Peter:
Abundance360Summit (www.A360.com)

Join Peter Diamandis's Abundance 360 Community—a group of entrepreneurs passionate about generating extraordinary wealth while creating a world of abundance.

Membership in the Abundance 360 Community is highly curated, focused on Top of Form entrepreneurs committed to expanding their business 10x to 100x, on a global level. The summit and monthly webinars are personally taught and curated by Diamandis, in cooperation with Singularity University and Strategic Coach. The content focuses on making exponential tools and technologies teachable and immediately usable in your business and your life. Diamandis has committed to running the Abundance 360 summit for the next twenty-five years. In a world of rapid, unpredictable growth, this is a program to count on every January to give you a road map for the year. To join the Community and begin receiving personal coaching from Peter, please visit www.A360.com and start the application process.

Flow Genome Project:
Trainings and Programs

At the Flow Genome Project, the most consistent question we get asked after people learn about flow's extraordinary impact on peak performance is how individuals and organizations can get more of it. If you want to learn more about how flow can help seriously raise your game, the Flow Genome Project offers a variety of courses and training programs, ranging from individual online trainings to multiday or multimonth corporate trainings. Details about all of our programs are available at www.FlowGenomeProject.com.

Singularity University Courses and Programs

If you enjoyed learning about Singularity University (SU) and would like to participate in one of our programs, you are welcome to get involved. Graduate and postgraduate students can apply for the ten-week Graduate Studies Program (GSP). Others, including executives, investors, and entrepreneurs, can apply for the six-day Executive programs held on a regular basis at the SU Campus in Mountain View, California. Details on both programs are available at the university's website, www.SingularityU.org.

XPRIZE Foundation Membership

Philanthropists and corporate executives interested in the design or funding of an XPRIZE can learn more at www.xprize.org. Or for more information simply shoot us an email at alliances@xprize.org.

Keynotes: Hiring Diamandis and/or Kotler for Corporate/Association Events

For information about hiring Peter Diamandis to give a keynote address for your corporation or association, more information is available at www.Diamandis.com.

For information about hiring Steven Kotler to give a keynote address for your corporation or association, more information is available at www.StevenKotler.com.

Thank you for taking the time to read *Bold*. We hope our contrarian view of the future has provided an antidote to some of today's pessimisms. Creating abundance is humanity's grandest challenge—one that together, with intention and action, we can make happen within our lifetime.

ACKNOWLEDGMENTS

BOLD has benefited greatly from the generous wisdom of a great many people. First off, the authors would like to express deep gratitude to their families, Jet, Dax and Kristen Diamandis and Joy Nicholson Kotler, for their incredible patience and support. We'd also like to thank our agent, John Brockman, our editor Thomas LeBein, Brit Hvide and everyone at Simon & Schuster who worked so doggedly on this project.

For insight and feedback, we thank a group of experts and friends, including Salim Ismail, Marcus Shingles, Andrew Hessel, Michael Wharton, Jamie Whealand, and Fred MacDonald.

Special thanks goes to Peter's team at PHD Ventures (Marissa Brassfield, Cody Rapp, Maxx Bricklin, and Kelley Lujan) for their incredible support in doing research, crowdsourcing content, and providing twenty-four by seven input. And recognition to Connie Fox for the herculean task of coordinating Peter's schedule and life.

On the research and inspiration front, we are also grateful to the Singularity University family of alumni, faculty, and staff under the leadership of co-founder and chancellor Ray Kurzweil and CEO Rob Nail. We are also thankful to the XPRIZE Family for the support and inspiration in writing this book and their dedication to solving humanity's grand challenges. In particular Bob Weiss, Eileen Bartholomew, Trish Halamandaris, Paul Rappoport, Chris Frangione, and the entire XPRIZE staff.

For coaching and rock-star marketing advice we consider ourselves lucky to be mentored by Dan Sullivan (and the team at Strategic Coach), Joe Polish, Brendon Burchard, and Mike Cline.

Peter is thankful to the incredibly energizing and uplifting music of Sarah Brightman, which was listened to for countless hours while writing this book on airplanes and in hotel rooms.

Great appreciation goes to writer Bob Hughes for his support in authoring the early blogs to investigate the content of *BOLD*, and to Michael Drew for his support in promoting this book.

Finally, we wish to thank the hundreds and thousands of fans and readers who have given us feedback through Google+, Facebook, and email as we developed this content. The names of these friends can be found here: www.BoldtheBook.com/supporters.

NOTES

Introduction: Birth of the Exponential Entrepreneur

1 Mary Bagley, "Cretaceous Period: Facts About Animals, Plants & Climate," *LiveScience*, May 1, 2013.
2 Paul R. Renne et al., "Time Scales of Critical Events Around the Cretaceous-Paleogene Boundary," *Science* 339, no. 6120 (February 8, 2013): 684–687.
3 This has been documented all over the place, but if you want to see a really cool infographic comparing a typical smartphone to a 1985-era supercomputer, check out http://www.charliewhite.net/2013/09/smartphones-vs-super computers/.
4 Ray Kurzweil, "The Law of Accelerating Returns," *Kurzweil Accelerating Intelligence*, March 7, 2001, http://www.kurzweilai.net/the-law-of-accelerating -returns.
5 Abundance was named a top five book of the year by both *Fortune* and *Money*.

PART ONE: BOLD TECHNOLOGY
Chapter One: Good-bye Linear Thinking . . . Hello, Exponential

1 Elizabeth Brayer, *George Eastman: A Biography* (Baltimore, MD: The Johns Hopkins University Press, 1996), 24–72. Or see http://www.kodak.com/ek /US/en/Our_Company/History_of_Kodak/George_Eastman.htm.
2 See http://www.kodak.com/ek/US/en/Our_Company/History_of_Kodak /George_Eastman.htm.
3 Ibid.
4 See "Kodak Moments: Steve Sasson, Digital Camera Inventor," https://www .youtube.com/watch?v=wfnpVRiiwnM.

5   John Pavlus, "How Steve Sasson Invented The Digital Camera," *Fast Company*, April 12, 2011, http://www.fastcodesign.com/1663611/how-steve-sasson-invented-the-digital-camera-video.

6   Steve Sasson, "Disruptive Innovation: The Story of the First Digital Camera," Linda Hall Library Lectures, October 26, 2011.

7   Andrew Martin, "Negative Exposure for Kodak," *New York Times*, October 20, 2011, http://www.nytimes.com/2011/10/21/business/kodaks-bet-on-its-printers-fails-to-quell-the-doubters.html?pagewanted=all.

8   Pavlus, "How Steve Sasson Invented The Digital Camera."

9   Gordon E. Moore, "Cramming more components onto integrated circuits," *Electronics*, April 19, 1965, 4.

10 Ray Kurzweil, "The Law of Accelerating Returns."

11 Michael J. de la Merced, "Eastman Kodak Files for Bankruptcy," *New York Times*, January 19, 2012, http://dealbook.nytimes.com/2012/01/19/eastman-kodak-files-for-bankruptcy/.

12 Chris Anderson, *Free: How Today's Smartest Businesses Profit by Giving Something for Nothing* (New York: Hyperion, 2010), 2–3.

13 Elizabeth Palmero, "Google Invests Billions on Satellites to Expand Internet Access," *Scientific American*, June 5, 2014.

14 Richard Foster and Sarah Kaplan, *Creative Destruction: Why Companies That Are Built to Last Underperform the Market—And How to Successfully Transform Them* (New York: Crown Business, 2001), 8.

15 Babson Olin School of Business Advertisement, *Fast Company*, April 2011, 121.

16 Foster and Kaplan, *Creative Destruction*.

17 Salim Ismail, AI, 2012.

18 Virginia Heffernan, "How We All Learned to Speak Instagram," *Wired*, April 2013, http://www.wired.com/2013/04/instagram-2/.

19 Chenda Ngak, "Instagram for Android gets 1 million downloads in first day," CBS News, April 4, 2012, http://www.cbsnews.com/news/instagram-for-android-gets-1-million-downloads-in-first-day/.

20 Joanna Stern, "Facebook Buys Instagram for $1 Billion," ABC News, April 27, 2012, http://abcnews.go.com/blogs/technology/2012/04/facebook-buys-instagram-for-1-billion/.

21 Unless otherwise noted, all Ben Kauffman information comes from a series of AIs conducted in 2013.

22 Issie Lapowsky, "Quirky Lands $79 Million in Funding," *Inc.*, November 13, 2013, http://www.inc.com/issie-lapowsky/quirky-79-million-funding-connected-home.html.

23 AI conducted 2013.

24 Austin Carr, "Inside Airbnb's Grand Hotel Plans," *Fast Company*, April 2014.

25 Serena Saitto and Brad Stone, "Uber Sets Valuation Record of $17 Billion in New Funding," Bloomberg.com, January 7, 2014. See http://www.bloomberg .com/news/2014-06-06/uber-sets-valuation-record-of-17-billion-in-new-funding .html.

Chapter Two: Exponential Technology: The Democratization of the Power to Change the World

1 Edwin A. Locke, *The Prime Movers* (New York: AMACOM, 2000).
2 AI conducted 2013.
3 Steven Kotler, "The Whole Earth Effect," *Plenty*, no. 24 (October/November 2008): 84–91.
4 J. C. R Licklider, "Memorandum For Members and Affiliates of the Intergalactic Computer Network," Advanced Research Projects Agency, April 23, 1963. See http://www.kurzweilai.net/memorandum-for-members-and-affiliates-of -the-intergalactic-computer-network.
5 Chris Anderson, "The Man Who Makes the Future: *Wired* Icon Marc Andreessen," *Wired*, April 24, 2012, http://www.wired.com/2012/04/ff_andreessen /all/.
6 Ian Peter, "History of the World Wide Web," Net History, http://www.net history.info/History%20of%20the%20Internet/web.html.
7 McKinsey Global Institute, "Manufacturing the future: The next era of global growth and innovation," McKinsey & Company, November 2012, http:// www.mckinsey.com/insights/manufacturing/the_future_of_manufacturing.
8 Institute of Human Origins, "Earliest Stone Tool Evidence Revealed," *Becoming Human*, August 11, 2010, http://www.becominghuman.org/node/news /earliest-stone-tool-evidence-revealed.
9 Pagan Kennedy, "Who Made That 3-D Printer," *New York Times Magazine*, November 22, 2013, http://www.nytimes.com/2013/11/24/magazine/who -made-that-3-d-printer.html.
10 In full disclosure, Peter Diamandis is a member of the 3D Systems Board of Directors.
11 All Avi Reichenthal quotes come from a series of AIs conducted between 2012 and 2014.
12 Based on approximate average share price for 2014.
13 AI, June 2014.
14 AI with Jay Rogers, 2014.
15 David Szondy, "SpaceX completes qualification test of 3D-printed Super-Draco thruster," *Gizmag*, May 28, 2014, http://www.gizmag.com/superdraco -test/32292/.

16 James Hagerty and Kate Linebaugh, "Next 3-D Frontier: Printed Plane Parts,"
    Wall Street Journal, July 14, 2012, http://online.wsj.com/news/articles/SB1000
    1424052702303933404577505080296858896.

17 Tim Catts, "GE Turns to 3D Printers for Plane Parts," *Bloomberg Business-
    week*, November 27, 2013, http://www.businessweek.com/articles/2013-11-27
    /general-electric-turns-to-3d-printers-for-plane-parts.

18 All quotes about Made In Space come from an AI with Michael Chen con-
    ducted 2013.

19 Brian Dodson, "Launch your own satellite for US $8000," *Gizmag*, April 22,
    2012, http://www.gizmag.com/tubesat-personal-satellite/22211/.

20 Statista, "Statistics and facts on the Toy Industry," Statista.com, 2012, http://
    www.statista.com/topics/1108/toy-industry/.

21 Unless otherwise noted, all Alice Taylor quotes and facts come from an AI con-
    ducted in 2013.

22 Cory Doctorow, *Makers* (New York: Tor Books, 2009).

Chapter Three: Five to Change the World

1   Adrian Kingsley-Hughes, "Mobile gadgets driving massive growth in touch sen-
    sors," *ZDNet*, June 18, 2013, http://www.zdnet.com/mobile-gadgets-driving
    -massive-growth-in-touch-sensors-7000016954/.

2   Peter Kelly-Detwiler, "Machine to Machine Connections—The Internet of
    Things—And Energy," *Forbes*, August 6, 2013, http://www.forbes.com/sites
    /peterdetwiler/2013/08/06/machine-to-machine-connections-the-internet-of
    -things-and-energy/.

3   See http://www.shotspotter.com.

4   Clive Thompson, "No Longer Vaporware: The Internet of Things Is Finally Talking,"
    Wired, December 6, 2012, http://www.wired.com/2012/12/20-12-st_thompson/.

5   Brad Templeton, "Cameras or Lasers?," *Templetons*, http://www.templetons.
    com/brad/robocars/cameras-lasers.html.

6   See http://en.wikipedia.org/wiki/Passenger_vehicles_in_the_United_States.

7   Commercial satellite players include: PlanetLabs (already launched), Sky-
    box (launched and acquired by Google), Urthecast (launched), and two still-
    confidential companies still under development (about which Peter Diamandis
    has firsthand knowledge).

8   Stanford University, "Need for a Trillion Sensors Roadmap," Tsensorsummit.
    org, 2013, http://www.tsensorssummit.org/Resources/Why%20TSensors%20
    Roadmap.pdf.

9   Rickie Fleming, "The battle of the G networks," NCDS.com blog, June 28,
    2014, http://www.ncds.com/ncds-business-technology-blog/the-battle-of-the
    -g-networks.

10 AI with Dan Hesse, 2013–14.

11 Unless otherwise noted, all IoT information and Padma Warrior quotes come from an AI with Padma, 2013.

12 Cisco, "2013 IoE Value Index," Cisco.com, 2013, http://internetofeverything .cisco.com/learn/2013-ioe-value-index-whitepaper.

13 NAVTEQ, "NAVTEQ Traffic Patterns," Navmart.com, 2008, http://www .navmart.com/pdf/NAVmart_TrafficPatterns.pdf.

14 Juho Erkheikki, "Nokia to Buy Navteq for $8.1 Billion, Take on TomTom (Update 7)," *Bloomberg*, October 1, 2007, http://www.bloomberg.com/apps /news?pid=newsarchive&sid=ayyeY1gIHSSg.

15 John Swartz, "Show me the Waze: Google maps a $1 billion deal," *USA Today*, June 12, 2013, http://www.usatoday.com/story/tech/2013/06/11/google -waze/2411871/.

16 Cisco, "2013 IoE Value Index."

17 See http://www.getturnstyle.com.

18 See http://www.adheretech.com.

19 See http://www.coherohealth.com/#home.

20 AI with Briggs conducted 2014.

21 J. P. Mangalindan, "A digital maestro for every object in the home," *Fortune*, June 7, 2013, http://fortune.com/2013/06/07/a-digital-maestro-for-every -object-in-the-home/.

22 Unless otherwise noted, all Bass quotes and Autodesk information comes from a series of AIs with Carl Bass conducted 2012–2014.

23 Michio Kaku, "The Future of Computing Power [Fast, Cheap, and Invisible]," *Big Think*, April 24, 2010, http://bigthink.com/dr-kakus-universe/the-future -of-computing-power-fast-cheap-and-invisible.

24 AI with Graham Weston, 2013.

25 *2001: A Space Odyssey*, directed by Stanley Kubrick (1968; Beverly Hills, CA: Metro-Goldwyn-Mayer), DVD release, 2011.

26 *Iron Man*, directed by Jon Favreau (2008; Burbank, CA: Walt Disney Studios), DVD.

27 AI with Ray Kurzweil, 2013.

28 See: http://www.xprize.org/ted. As of the end of 2014, this prize is only in con- cept form. A detailed design and a design sponsor is still required.

29 "Ray Kurzweil: The Coming Singularity, Your Brain Year 2029," *Big Think*, June 22, 2013, https://www.youtube.com/watch?v=6adugDEmqBk.

30 John Ward, "The Services Sector: How Best To Measure It?," *International Trade Administration*, October 2010, http://trade.gov/publications/ita-newsletter /1010/services-sector-how-best-to-measure-it.asp.

31 AI with Jeremy Howard, 2013.

32 For information on the German Traffic Sign Recognition Benchmark see http://benchmark.ini.rub.de.

33 Geoffrey Hinton et al., "ImageNet Classification with Deep Convolutional Neural Networks," http://www.cs.toronto.edu/~fritz/absps/imagenet.pdf.

34 John Markoff, "Armies of Expensive Lawyers, Replaced By Cheaper Software," *New York Times*, March 4, 2011, http://www.nytimes.com/2011/03/05/science/05legal.html?pagewanted=all.

35 David Schatsky and Vikram Mahidhar, "Intelligent automation: A new era of innovation," Deloitte University Press, January 22, 2014, http://dupress.com/articles/intelligent-automation-a-new-era-of-innovation/.

36 John Markoff, "Computer Wins on 'Jeopardy!': Trivial, It's Not," *New York Times*, February 16, 2011, http://www.nytimes.com/2011/02/17/science/17jeopardy-watson.html?pagewanted=all.

37 "IBM Watson's Next Venture: Fueling New Era of Cognitive Apps Built in the Cloud by Developers," IBM Press Release, November 14, 2013, http://www-03.ibm.com/press/us/en/pressrelease/42451.wss.

38 Nancy Dahlberg, "Modernizing Medicine, supercomputer Watson partner up," *Miami Herald*, May 16, 2014.

39 AI with Daniel Cane, 2014.

40 Ray Kurzweil, "The Law of Accelerating Returns."

41 Daniela Hernandez, "Meet the Man Google Hired to Make AI a Reality," *Wired*, January 2014, http://www.wired.com/2014/01/geoffrey-hinton-deep-learning/.

42 AI with Geordie Rose, 2014.

43 See http://1qbit.com.

44 John McCarthy, Marvin Minsky, Nathaniel Rochester, and Claude E. Shannon, "A Proposal for the Dartmouth Summer Research Project on Artificial Intelligence," *AI Magazine*, August 31, 1955, 12–14.

45 Jim Lewis, "Robots of Arabia," *Wired*, Issue 13.11 (November 2005).

46 Garry Mathiason et al., "The Transformation of the Workplace Through Robotics, Artificial Intelligence, and Automation," *The Littler Report*, February 2014, http://documents.jdsupra.com/d4936b1e-ca6c-4ce9-9e83-07906bfca22c.pdf.

47 See http://www.rethinkrobotics.com.

48 All Dan Barry quotes in this section come from an AI conducted 2013.

49 The Cambrian explosion was an evolutionary event beginning about 542 million years ago, during which most of the major animal phyla appeared.

50 See "Amazon Prime Air," Amazon.com, http://www.amazon.com/b?node=8037720011.

51 Jonathan Berr, "Google Buys 8 Robotics Companies in 6 Months: Why?" CBSnews.com, *CBS Money Watch*, December 16, 2013, http://www.cbsnews.com/news/google-buys-8-robotics-companies-in-6-months-why/.

52 Brad Stone, "Smarter Robots, With No Wage Demands," *Bloomberg Businessweek*, September 18, 2012, http://www.businessweek.com/articles/2012-09-18/smarter-robots-with-no-pesky-uprisings.

53 Aviva Hope Rutkin, "Report Suggests Nearly Half of U.S. Jobs Are Vulnerable to Computerization," *MIT Technology Review*, September 12, 2013, http://www.technologyreview.com/view/519241/report-suggests-nearly-half-of-us-jobs-are-vulnerable-to-computerization/.

54 Lee Chyen Yee and Jim Clare, "Foxconn to rely more on robots; could use 1 million in 3 years," *Reuters*, August 1, 2011, http://www.reuters.com/article/2011/08/01/us-foxconn-robots-idUSTRE77016B20110801.

55 Jennifer Wang, "Cutting-Edge Startups Leading the Robotic Revolution," *Entrepreneur*, June 3, 2013, http://www.entrepreneur.com/article/226397.

56 For a great introduction to and overview of synthetic biology, see Andrew Hessel's blog at www.andrewhessel.com.

57 Elsa Wenzel, "Scientists create glow-in-the-dark cats," CNET, December 12, 2007, http://www.cnet.com/news/scientists-create-glow-in-the-dark-cats/.

58 All Andrew Hessel quotes come from a series of AIs conducted in 2013.

59 For a full recounting of Venter's discoveries, see Steven Kotler, Marc Goodman, and Andrew Hessel, "Hacking the President's DNA," *The Atlantic*, October 24, 2012, http://www.theatlantic.com/magazine/archive/2012/11/hacking-the-presidents-dna/309147/.

60 AI with Carlos Olguin conducted in 2013. Also see http://www.autodesk research.com/projects/cyborg.

61 See http://www.humanlongevity.com.

62 Walter Isaacson, *Steve Jobs* (New York: Simon & Schuster, 2011), 92.

PART TWO: BOLD MINDSET
Chapter Four: Climbing Mount Bold

1 "Skunk works," Worldwidewords.com, http://www.worldwidewords.org/qa/qa-sku1.htm.

2 Lockheed Martin, "Skunk Works Origin Story," Lockheedmartin.com, http://www.lockheedmartin.com/us/aeronautics/skunkworks/origin.html.

3 Matthew E May, "The Rules of Successful Skunk Works Projects," *Fast Company*, October 9, 2012, http://www.fastcompany.com/3001702/rules-successful-skunk-works-projects.

4 Unless otherwise noted, all Gary Latham and Edwin Locke quotes are taken from a series of AIs conducted in 2013.

5 Edwin Locke and Gary Latham, "New Directions in Goal-Setting Theory," *Current Directions in Psychological Science* 15, no. 5 (2006): 265–68.

6 Lockheed Martin, "Kelly's 14 Rules & Practices," Lockheedmartin.com, http://www.lockheedmartin.com/us/aeronautics/skunkworks/14rules.html.

7 Jeff Bezos, "2012 re: Invent Day 2: Fireside Chat with Jeff Bezos & Werner Vogels," November 29, 2012. See https://www.youtube.com/watch?v=O4MtQGRIIuA.

8 Dominic Basulto, "The new #Fail: Fail fast, fail early and fail often," *Washington Post*, May 30, 2012, http://www.washingtonpost.com/blogs/innovations/post/the-new-fail-fail-fast-fail-early-and-fail-often/2012/05/30/gJQAKA891U_blog.html.

9 John Anderson, "Change on a Dime: Agile Design," *UX Magazine*, July 19, 2011, http://uxmag.com/articles/change-on-a-dime-agile-design.

10 AI with Ismail, 2013.

11 For an amazing breakdown of these ideas, see Dan Pink, "RSA Animate—Drive: The surprising truth about what motivates us," RSA, April 1, 2010, https://www.youtube.com/watch?v=u6XAPnuFjJc.

12 Daniel Kahneman, "The riddle of experience vs. memory," TED, March 1, 2010, http://www.ted.com/talks/daniel_kahneman_the_riddle_of_experience_vs_memory.

13 Daniel H. Pink, *Drive: The Surprising Truth About What Motivates Us* (New York: Riverhead Books, 2010).

14 Christopher Mims, "When 110% won't do: Google engineers insist 20% time is not dead—it's just turned into 120% time," qz.com, August 16, 2013.

15 James Marshall Reilly, "The Zappos Story: How Failure can Fuel Business Success," Monster.com, http://hiring.monster.com/hr/hr-best-practices/workforce-management/hr-management-skills/business-success.aspx.

16 All Astro Teller quotes come from a series of AIs conducted between 2013 and 2014.

17 Susan Wojcicki, "The Eight Pillars of Innovation," thinkwithgoogle.com, July 2011, http://www.thinkwithgoogle.com/articles/8-pillars-of-innovation.html.

18 For a much deeper look at flow and its impact on performance see Steven Kotler, *The Rise of Superman: Decoding the Science of Ultimate Human Performance* (New York: New Harvest, 2014).

19 AI with John Hagel conducted 2014.

20 Steven Kotler and Jamie Wheal, "Five Surprising Ways Richard Branson Harnessed Flow to Build A Multi-Billion Dollar Empire," *Forbes*, March 25, 2014, http://www.forbes.com/sites/stevenkotler/2014/03/25/five-surprising-ways-richard-branson-harnessed-flow-to-build-a-multi-billion-dollar-empire/.

21 Steven Kotler, "The Rise of Superman: 17 Flow Triggers," Slideshare.net, March 2014, http://www.slideshare.net/StevenKotler/17-flow-triggers.

22 AI with Ned Hallowell conducted 2013.

23 Kevin Rathunde, "Montessori Education and Optimal Experience: A Framework for New Research," *The NAMTA Journal* (Winter 2001): 11–43.

24 Mihaly Csikszentmihalyi, *Flow: The Psychology of Optimal Experience* (New York: Harper & Row, 1990), 48–70.

25 For a great breakdown of group flow and the social triggers see Keith Sawyer, *Group Genius: The Creative Power of Collaboration* (New York: Basic Books), 2008.

26 AI with Ismail, 2013.

Chapter Five: The Secrets of Going Big

1 Jon Stewart, *The Daily Show*, April 24, 2012.
2 See http://www.planetaryresources.com.
3 Space Adventures was cofounded in partnership with Mike McDowell, who served as the company's first CEO and chairman.
4 For a nice breakdown of Branson's space plans, see Elizabeth Howell, "Virgin Galactic: Richard Branson's Space Tourism Company," Space.com, December 20, 2012, http://www.space.com/18993-virgin-galactic.html.
5 See http://www.blueorigin.com.
6 Justine Bachman, "Elon Musk Wants SpaceX to Replace Russia as NASA's Space Station Transport," *Bloomberg Businessweek,* April 30, 2014, http://www.businessweek.com/articles/2014-04-30/elon-musk-wants-spacex-to-replace-russia-as-nasas-space-station-transport.
7 AI with Chris Anderson conducted 2013.
8 Mikhail S. Arlazorov, "Konstantin Eduardovich Tsiolkovsky," *Britannica.com,* May 30, 2013, http://www.britannica.com/EBchecked/topic/607781/Konstantin-Eduardovich-Tsiolkovsky.
9 For a breakdown of these missions, see Steven Kotler, "The Great Galactic Gold Rush," *Playboy,* March, 2011, available at: www.stevenkotler.com.
10 See http://seds.org.
11 The early founding days of SEDS was supported by a number of key individuals who deserve recognition, including James Muncy, Morris Hornik, Maryann Grams, Frank Taylor, Brian Ceccarelli, Eric Dahlstrom, David C. Webb, Gregg Maryniak, Scott Scharfman, and Eric Drexler.
12 See http://www.isunet.edu.
13 The event was called Space Fair. It was held in 1983, 1985, and 1987. Ken Sunshine, vice-chairman, deserves recognition. Special thanks to MIT president Paul Gray, who supported me far beyond what I deserved and taught me that MIT has viscosity. You get things done as long as you keep pushing.
14 Locke and Latham, "New Directions in Goal-Setting Theory."
15 It was my dear friend Gregg Maryniak who first introduced me to this story. As it has been fundamental to my success, a deep debt of gratitude is owed.
16 Wikipedia does a great job with the history of "stone soup," see http://en.wikipedia.org/wiki/Stone_Soup. Also see Marcia Brown, *Stone Soup* (New York: Aladdin Picture Books), 1997.
17 AI with Hagel.
18 John Hagel, "Pursuing Passion," *Edge Perspectives with John Hagel,* November 14, 2009, http://edgeperspectives.typepad.com/edge_perspectives/2009/11/pursuing-passion.html.
19 Gregory Berns, "In Hard Times, Fear Can Impair Decision Making," *New York Times,* December 6, 2008.

Chapter Six: Billionaire Wisdom: Thinking at Scale

1 Elon Musk, "The Rocket Scientist Model for *Iron Man*," *Time*, http://content
   .time.com/time/video/player/0,32068,81836143001_1987904,00.html.

2 Unless otherwise noted, historical details and Musk quotes come from a series
   of AIs between 2012 and 2014.

3 AI, XPRIZE Adventure Trip, February 2013.

4 Thomas Owen, "Tesla's Elon Musk: 'I Ran Out of Cash,' " *VentureBeat*, May
   2010, http://venturebeat.com/2010/05/27/elon-musk-personal-finances/.

5 Andrew Sorkin, Dealbook: "Elon Musk, of PayPal and Tesla Fame, Is Broke,"
   New York Times, June 2010, http://dealbook.nytimes.com/2010/06/22/sorkin
   -elon-musk-of-paypal-and-tesla-fame-is-broke/?_php=true&_type=blogs&_r=0.

6 SpaceX, "About Page," http://www.spacex.com/about.

7 Kenneth Chang, "First Private Craft Docks With Space Station," *New York
   Times*, May 2012, http://www.nytimes.com/2012/05/26/science/space/space
   -x-capsule-docks-at-space-station.html.

8 Elon Musk interviewed by Kevin Fong, *Scott's Legacy*, a BBC Radio 4 program,
   cited in Jonathan Amos, "Mars for the 'average person,' " BBC News, March
   20, 2012, http://www.bbc.com/news/health-17439490.

9 Diarmuid O'Connell, Statement from Tesla's vice president of corporate and
   business development, reported in Hunter Walker, "White House Won't Back
   Tesla in Direct Sales Fight" in *Business Insider*, July 14, 2014, http://www
   .businessinsider.com/white-house-wont-back-tesla-2014-7.

10 Daniel Gross, "Elon's Élan," *Slate*, April 30, 2014, http://www.slate.com
   /articles/business/moneybox/2014/04/tesla_and_spacex_founder_elon_musk_
   has_a_knack_for_getting_others_to_fund.html.

11 Kevin Rose, "Elon Musk," Video Interview, Episode 20, *Foundation*, September 2012, http://foundation.bz/20/.

12 Daniel Kahneman, "Why We Make Bad Decisions About Money (And What
   We Can Do About It)," *Big Think*, Interview, June 2013, http://bigthink
   .com/videos/why-we-make-bad-decisions-about-money-and-what-we-can-do
   -about-it-2.

13 Chris Anderson, "The Shared Genius of Elon Musk and Steve Jobs", *Fortune*,
   November 21, 2013, http://fortune.com/2013/11/21/the-shared-genius-of
   -elon-musk-and-steve-jobs/.

14 AI, September 2013.

15 Eric Kelsey, "Branson recalls tears, $1 billion check in Virgin Records sale,"
   Reuters, October 23, 2013, http://www.reuters.com/article/2013/10/24/us
   -richardbranson-virgin-idUSBRE99N01U20131024.

16 *Forbes*, The World's Billionaires: #303 Richard Branson, August 2014, http://
   www.forbes.com/profile/richard-branson/.

17 Richard Branson, "BA Can't Get It Up - best stunt ever?," Virgin, 2012, http://www.virgin.com/richard-branson/ba-cant-get-it-up-best-stunt-ever.

18 Richard Branson, *Screw It, Let's Do It: Lessons in Life* (Virgin Books, March 2006).

19 "Galactic Announces Partnership," Virgin Galactic, July 2009, http://www.virgingalactic.com/news/item/galactic-anounces-partnership/.

20 Nour Malas, "Abu Dhabi's Aabar boosts Virgin Galactic stake," Market Watch, October 19, 2011, http://www.marketwatch.com/story/abu-dhabis-aabar-boosts-virgin-galactic-stake-2011-10-19.

21 Loretta Hidalgo Whitesides, "Google and Virgin Team Up to Spell 'Virgle,'" *Wired*, April 1, 2008, http://www.wired.com/2008/04/google-and-virg/.

22 "Jeffrey Preston Bezos," *Bio*. A&E Television Networks, 2014, http://www.biography.com/people/jeff-bezos-9542209.

23 Brad Stone, *The Everything Store: Jeff Bezos and the Age of Amazon* (New York: Little, Brown, 2014).

24 "Jeffrey P. Bezos Biography," Academy of Achievement, November 2013, http://www.achievement.org/autodoc/page/bez0bio-1.

25 Suzanne Galante and Dawn Kawamoto, "Amazon IPO skyrockets," CNET, May 15, 1997, http://news.cnet.com/2100-1001-279781.html.

26 Jeffery P. Bezos, "1997 Letter to Shareholders," Amazon.com, *Ben's Blog*, 1997, http://benhorowitz.files.wordpress.com/2010/05/amzn_shareholder-letter-20072.pdf

27 "2012 re: Invent Day 2: Fireside Chat with Jeff Bezos and Werner Vogels."

28 Julie Bort, "Amazon Is Crushing IBM, Microsoft, And Google in Cloud Computing, Says Report," *Business Insider*, November 26, 2013, http://www.businessinsider.com/amazon-cloud-beats-ibm-microsoft-google-2013-11#ixzz37zMH8gUr.

29 James Stewart, "Amazon Says Long Term and Means It," *New York Times*, December 16, 2011, http://www.nytimes.com/2011/12/17/business/at-amazon-jeff-bezos-talks-long-term-and-means-it.html?pagewanted=all&_r=0.

30 "Utah Technology Council Hall of Fame—Jeff Bezos Keynote," Utah Technology Council, published online April 30, 2013, https://www.youtube.com/watch?v=G-0KJF3uLP8.

31 "About Blue Origin," Blue Origin, July 2014, http://www.blueorigin.com/about/.

32 Alistair Barr, "Amazon testing delivery by drone, CEO Bezos Says," *USA Today*, December 2, 2013, referencing a *60 Minutes* interview with Jeff Bezos, http://www.usatoday.com/story/tech/2013/12/01/amazon-bezos-drone-delivery/3799021/.

33 Jay Yarow, "Jeff Bezos' Shareholder Letter Is Out," *Business Insider*, April 10, 2014, http://www.businessinsider.com/jeff-bezos-shareholder-letter-2014-4.

34 "Larry Page Biography," Academy of Achievement, January 21, 2011, http://www.achievement.org/autodoc/page/pag0bio-1.

35 Marcus Wohlsen, "Google Without Larry Page Would Not Be Like Apple Without Steve Jobs," *Wired*, October 18, 2013, http://www.wired.com/2013/10/google-without-page/.

36 Google Inc., 2012, Form 10-K 2012, retrieved from SEC Edgar website: http://www.sec.gov/Archives/edgar/data/1288776/000119312513028362/d452134d10k.htm.

37 Larry Page, "Beyond Today—Larry Page—Zeitgeist 2012," Google Zeitgeist, Zeitgeist Minds, May 22, 2012, https://www.youtube.com/watch?v=Y0WH-CoFwn4.

38 Matt Ridley, *The Rational Optimist: How Prosperity Evolves* (New York: Harper-Collins, 2010).

39 Larry Page, "Google I/O 2013: Keynote," Google I/O 2013, Google Developers, May 15, 2013, https://www.youtube.com/watch?v=9pmPa_KxsAM.

40 Joann Muller, "No Hands, No Feet: My Unnerving Ride in Google's Driverless Car," *Forbes*, March 21, 2013, http://www.forbes.com/sites/joannmuller/2013/03/21/no-hands-no-feet-my-unnerving-ride-in-googles-driverless-car/.

41 Robert Hof, "10 Breakthrough Technologies 2013: Deep Learning," *MIT Technology Review*, April 23, 2013, http://www.technologyreview.com/featuredstory/513696/deep-learning/.

42 Steven Levy, "Google's Larry Page on Why Moon Shots Matter," *Wired*, January 17, 2013, http://www.wired.com/2013/01/ff-qa-larry-page/all/.

43 Larry Page, "Beyond Today—Larry Page—Zeitgeist 2012."

44 Larry Page, "Google+: Calico Announcement," Google+, September 2013, https://plus.google.com/+LarryPage/posts/Lh8SKC6sED1.

45 Harry McCracken and Lev Grossman, "Google vs. Death," *Time*, September 30, 2013, http://time.com/574/google-vs-death/.

46 Jason Calacanis, "#googlewinseverything (part 1)," *Launch*, October 30, 2013, http://blog.launch.co/blog/googlewinseverything-part-1.html.

PART THREE: THE BOLD CROWD
Chapter Seven: Crowdsourcing: Marketplace of the Rising Billion

1 Netcraft Web Server Survey, Netcraft, Accessed June 2014, http://news.netcraft.com/archives/category/web-server-survey/.

2 AI with Jake Nickell and Jacob DeHart.

3 Jeff Howe, "The Rise of Crowdfunding," *Wired*, 2006, http://archive.wired.com/wired/archive/14.06/crowds_pr.html.

4 Rob Hof, "Second Life's First Millionaire," *Bloomberg Businessweek*, November 26, 2006, http://www.businessweek.com/the_thread/techbeat/archives /2006/11/second_lifes_fi.html.

5 Jeff Howe, "Crowdsourcing: A Definition," *Crowdsourcing*, http://crowdsourcing .typepad.com/cs/2006/06/crowdsourcing_a.html.

6 "Statistics," Kiva, http://www.kiva.org/about/stats.

7 Rob Walker, "The Trivialities and Transcendence of Kickstarter," *New York Times*, August 5, 2011, http://www.nytimes.com/2011/08/07/magazine/the -trivialities-and-transcendence-of-kickstarter.html?pagewanted=all&_r=0.

8 "Stats," Kickstarter, https://www.kickstarter.com/help/stats.

9 Doug Gross, "Google boss: Entire world will be online by 2020," CNN, April 15, 2013, http://www.cnn.com/2013/04/15/tech/web/eric-schmidt-internet/.

10 "Global entertainment and media outlook 2013–2017," Pricewaterhouse-Coopers, 2013, https://www.pwc.com/gx/en/global-entertainment-media -outlook/.

11 Freelancer Case Study based on a series of AIs.

12 Quoted from AI: Matt Barrie.

13 Tongal Case Study based on a series of AIs with James DeJulio.

14 reCAPTCHA and Duolingo Case Study based on a series of AIs with Luis von Ahn.

15 During the completion of this book, a Bay Area startup called Vicarious wrote an AI program able to solve (i.e., read) CAPTCHAs with an accuracy of 90 percent. As mentioned earlier, "crowdsourcing" is an interim solution until such AI comes fully online. This is a relevant example of that point.

16 "FAQ—Overview," Amazon Mechanical Turk, Amazon.com, Inc., 2014, https://www.mturk.com/mturk/help?helpPage=overview.

17 "What is Fiverr?," Fiverr.com, 2014, http://support.fiverr.com/hc/en-us /articles/201500776-What-is-Fiverr-.

18 Unless otherwise noted, all Matt Barrie quotes come from a 2013 AI.

19 AIs with Marcus Shingles, 2013–2014.

20 AI with Andrew Vaz.

21 "About Us," Freelancer.com, 2014, https://www.freelancer.com/info/about .php.

22 AI with Barrie.

23 Ibid.

24 AI with James DeJulio, 2013.

25 AI with Barrie.

26 Ibid.

27 "Vicarious AI passes first Turing Test: CAPTCHA," Vicarious, October 27, 2013, http://news.vicarious.com/post/65316134613/vicarious-ai-passes-first -turing-test-captcha.

Chapter Eight: Crowdfunding: No Bucks, No Buck Rogers

1 "Statistics about Business Size (including Small Business) from the U.S. Census Bureau," Statistics of US Businesses, United States Census Bureau, 2007, https://www.census.gov/econ/smallbus.html.

2 "Statistics about Business Size (including Small Business) from the U.S. Census Bureau."

3 Devin Thorpe, "Why Crowdfunding Will Explode in 2013," *Forbes*, October 15, 2012, http://www.forbes.com/sites/devinthorpe/2012/10/15/get-ready-here-it-comes-crowdfunding-will-explode-in-2013/.

4 Victoria Silchenko, "Why Crowdfunding Is The Next Big Thing: Let's Talk Numbers," *Huffington Post*, October 22, 2012, http://www.huffingtonpost.com/victoria-silchenko/why-crowdfunding-is-the-n_b_1990230.html.

5 Laurie Kulikowski, "How Equity Crowdfunding Can Swell to a $300 Billion Industry," *TheStreet*, January 14, 2013, http://www.thestreet.com/story/11811196/1/how-equity-crowdfunding-can-swell-to-a-300-billion-industry.html.

6 "Floating Pool Project Is Fully Funded And New Yorkers Everywhere Should Celebrate," *Huffington Post*, July 12, 2013, http://www.huffingtonpost.com/2013/07/12/floating-pool-project-is-fully-funded_n_3587814.html.

7 AI with Joshua Klein, 2013.

8 Dan Leone, "Planetary Resources Raises $1.5M for Crowdfunded Space Telescope," Space.com, July 14, 2013, http://www.space.com/21953-planetary-resources-crowdfunded-space-telescope.html.

9 For a good breakdown of these rules, please see http://www.cfira.org.

10 AI with Chance Barnett, 2013.

11 This information sits on a banner across the top of their landing page: https://www.crowdfunder.com, our numbers were gathered in June 2014.

12 See http://blog.angel.co/post/59121578519/wow-uber.

13 Tomio Geron, "AngelList, With SecondMarket, Opens Deals to Small Investors for as Little as $1K," *Forbes*, December 19, 2012, http://www.forbes.com/sites/tomiogeron/2012/12/19/angellist-with-secondmarket-opens-deals-to-small-investors-for-as-little-as-1k/.

14 John McDermott, "Pebble 'Smartwatch' Funding Soars on Kickstarter," *Inc.*, April 20, 2012, http://www.inc.com/john-mcdermott/pebble-smartwatch-funding-sets-kickstarter-record.html.

15 Dara Kerr, "World's first public space telescope gets Kickstarter goal," CNET, July 1, 2013, http://www.cnet.com/news/worlds-first-public-space-telescope-gets-kickstarter-goal/.

16 McDermott, "Pebble 'Smartwatch' Funding Soars on Kickstarter."

17 See https://www.indiegogo.com/projects/let-s-build-a-goddamn-tesla-museum--5.

18 Kerr, "World's first public space telescope gets Kickstarter goal."
19 Cade Metz, "Facebook Buys VR Startup Oculus for $2 Billion," *Wired*, March 25, 2014, http://www.wired.com/2014/03/facebook-acquires-oculus/.
20 All Indiegogo stats come from AIs with founders Danae Ringelmann and Slava Rubin, conducted in 2013.
21 Ibid.
22 Ibid.
23 AI with Eric Migicovsky, 2013.
24 See www.brainyquote.com/quotes/quotes/a/abrahamlin109275.html.
25 Eric Gilbert and Tanushree Mitra, "The Language that Gets People to Give: Phrases that Predict Success on Kickstarter," *CSCW'14*, February 15, 2014, http://comp.social.gatech.edu/papers/cscw14.crowdfunding.mitra.pdf.
26 AI with Ringelmann and Rubin, 2013.
27 AI with Migicovsky.
28 AI with Ringelmann and Rubin.
29 AI with Migicovsky.

Chapter Nine: Building Communities

1 Clay Shirky, "How cognitive surplus will change the world," TED, June 2010, https://www.ted.com/talks/clay_shirky_how_cognitive_surplus_will_change_the_world.
2 The term MTP was first described by Salim Ismail in his recent book *Exponential Organizations: Why new organizations are ten times better, faster, cheaper than yours (and what to do about it)* to describe a unique, powerful, and simple statement of your mission. Google's MTP is to "Organize the world's information." TED's MTP is "ideas worth spreading."
3 This is sometimes called Joy's Law, http://en.wikipedia.org/wiki/Joy's_Law_(management).
4 For DIY drones, see Chris Anderson, "How I Accidentally Kickstarted the Domestic Drone Boom," *Wired*, June 22, 2012, http://www.wired.com/2012/06/ff_drones. For Local Motors, localmotors.com.
5 AI with Gina Bianchini, 2014.
6 Joshua Klein, *Reputation Economics: Why Who You Know Is Worth More Than What You Have* (New York: Palgrave Macmillan Trade, 2013).
7 All Klein quotes come from an AI with Joshua Klein, 2014.
8 AI with Bianchini.
9 James Glanz, "What Else Lurks Out There? New Census of the Heavens Aims to Find Out," *New York Times*, March 17, 1998, http://www.nytimes.com/1998/03/17/science/what-else-lurks-out-there-new-census-of-the-heavens-aims-to-find-out.html.

10 All Kevin Schawinski quotes come from an AI conducted in 2013.

11 For a breakdown of Galaxy Zoo's history, see http://www.galaxyzoo.org/#/story.

12 All Jay Rogers quotes come from a series of AIs conducted in 2013 and 2014.

13 Reena Jana, "Local Motors: A New Kind of Car Company," *Bloomberg Businessweek*, November 3, 2009, http://www.businessweek.com/innovate/content/oct2009/id20091028_848755.htm.

14 Bureau of Labor Statistics, "Unemployment Rate for the 50 Largest Cities," United States Department of Labor, April 18, 2014, http://www.bls.gov/lau/lacilg10.htm.

15 AI with Bianchini.

16 Chris Anderson, "In the Next Industrial Revolution, Atoms Are the New Bits," *Wired*, January 25, 2010, http://www.wired.com/2010/01/ff_newrevolution/all/.

17 Megan Wollerton, "GE and Local Motors team up to make small-batch appliances," CNET, March 21, 2014, http://www.cnet.com/news/ge-and-local-motors-team-up-to-make-small-batch-appliances/.

18 All Jack Hughes quotes come from an AI conducted in 2014.

19 For a breakdown of the TopCoder rating system, see http://community.topcoder.com/longcontest/?module=Static&d1=support&d2=ratings.

20 Carolyn Johnson, "Thorny research problems, solved by crowdsourcing," *Boston Globe*, February 11, 2013, http://www.bostonglobe.com/business/2013/02/11/crowdsourcing-innovation-harvard-study-suggests-prizes-can-spur-scientific-problem-solving/JxDkOkuIKboRjWAoJpM0OK/story.html.

21 AI with Narinder Singh, 2014.

22 AI with Chris Anderson, 2013.

23 Richard Millington, "7 Contrary Truths About Online Communities," Feverbee.com, September 22, 2010, http://www.feverbee.com/2010/09/7truths.html.

24 AI with Jono Bacon, 2014.

25 Jolie O'Dell, "10 Fresh Tips for Community Managers," *Mashable*, April 13, 2010, http://mashable.com/2010/04/13/community-manager-tips/.

26 Seth Godin, "Why You Need to Lead A Tribe," Mixergy.com, January 13, 2009, http://mixergy.com/interviews/tribes-seth/.

27 AI with Better Blocks founder Jason Roberts, 2014. Also see his pretty amazing TEDx Talk: TEDxOU—Jason Roberts—How to Build a Better Block, https://www.youtube.com/watch?v=ntwqVDzdqAU.

Chapter Ten: Incentive Competitions: Getting the Best and Brightest to Help Solve Your Challenges

1 Charles Lindbergh, *The Spirit of St. Louis* (New York: Scribner, 1953).

2 Stephen Schaber, "Why Napoleon Offered a Prize for Inventing Canned

Food," *NPR*, March 5, 2012, http://www.npr.org/blogs/money/2012/03/01
/147751097/why-napoleon-offered-a-prize-for-inventing-canned-food.

3 Knowledge Ecology International, "Selected Innovation Prizes and Reward Programs," *KEI Research Note 2008:1*, http://keionline.org/misc-docs/research_
notes/kei_rn_2008_1.pdf.

4 All Marcus Shingles quotes come from an AI conducted 2014.

5 Burt Rutan, "The real future of space exploration," TED, February 2006,
https://www.ted.com/talks/burt_rutan_sees_the_future_of_space.

6 Statista, "Statistics and facts on Sports Sponsorship," *Sports sponsorship Statista
Dossier 2013*, March 2013.

7 Alice Roberts, "A true sea shanty: the story behind the Longitude prize," *The
Observer*, May 17, 2014, http://www.theguardian.com/science/2014/may/18
/true-sea-shanty-story-behind-longitude-prize-john-harrison.

8 See http://www.interculturalstudies.org/faq.html#quote.

9 Dan Heath and Chip Heath, "Get Back in the Box: How Constraints Can
Free Your Team's Thinking," *Fast Company*, December 1, 2007, http://www.fast
company.com/61175/get-back-box.

10 For all things LunarX, see http://www.googlelunarxprize.org.

11 Campbell Robertson and Clifford Krauss, "Gulf Spill Is the Largest of Its
Kind, Scientists Say," *New York Times*, August 2, 2010, http://www.nytimes
.com/2010/08/03/us/03spill.html?_r=0.

12 See http://www.iprizecleanoceans.org.

13 Special thanks to Cristin Dorgelo, then head of this XPRIZE, who spent her
summer on the hot and humid Jersey Shore overseeing the judging and operations.

14 See http://www.qualcommtricorderxprize.org.

15 Kate Greene, "The $1 Million Netflix Challenge," *MIT Technology Review*,
October 6, 2006, http://www.technologyreview.com/news/406637/the-1
-million-netflix-challenge/.

16 Jordan Ellenberg, "This Psychologist Might Outsmart the Math Brains Competing for the Netflix Prize," *Wired*, February 25, 2008, http://archive.wired
.com/techbiz/media/magazine/16-03/mf_netflix?currentPage=all.

17 Ibid.

18 See https://herox.com.

19 Special recognition to Emily Fowler, who helped to found HeroX and transi-
tioned from XPRIZE employment to VP of Possibilities at HeroX.

20 AI with Graham Weston, 2014.

21 "Geekdom and HeroX Launch San Antonio Entrepreneurial Exchange Challenge with $500,000 Prize," *Reuters*, January 16, 2014, http://www.reuters
.com/article/2014/01/16/idUSnMKWN1dCba+1e4+MKW20140116.

22 Vivek Wadhwa, "The powerful role of incentive competitions to spur innovation," *Washington Post*, May 21, 2014, http://www.washingtonpost.com/blogs

/innovations/wp/2014/05/21/the-powerful-role-of-incentive-competitions-to
-spur-innovation/.

23 Paul Wahl, "The Winner," *Popular Science*, January 1958.

24 Richard P. Feynman, "Plenty of Room at the Bottom," California Institute of
Technology, December 1959. For full transcript of the lecture, see http://www
.its.caltech.edu/~feynman/plenty.html.

25 Richard E. Smalley, "Dr. Feynman's Small Idea," *Innovation* 5, no. 5 (Octo-
ber/November 2007), http://www.innovation-america.org/dr-feynmans-small
-idea.

26 AI with Shingles.

27 For arguably the best bit of writing on the adjacent possible, see Steven Johnson, "The Genius of the Tinkerer," *Wall Street Journal*, September 25, 2010,
http://online.wsj.com/articles/SB10001424052748703989304575503730l0
1860838. Also see "The Adjacent Possible: A Talk with Stuart A. Kauffman,"
Edge.org, November 9, 2003, http://edge.org/conversation/the-adjacent
-possible.

INDEX

Italic numbers refer to charts/graphs

312   INDEX

Mycoskie, Blake, 80
Mycroft, Frank, 180
MySQL, 163

Napoléon I, Emperor of France, 245
Napster, 11
Narrative Science, 56
narrow framing, 121
NASA, 96, 97, 100, 102, 110, 123,
    221, 228, 244
  Ames Research Center of, 58
  Jet Propulsion Laboratory (JPL) of, 99
  Mars missions of, 99, 118
National Collegiate Athletic Association
    (NCAA), 226
National Institutes of Health, 64, 227
National Press Club, 251
navigation, in online communities, 232
Navteq, 47
Navy Department, US, 72
NEAR Shoemaker mission, 97
Netflix, 254, 255
Netflix Prize, 254–56
Netscape, 117, 143
networks and sensors, x, 14, 21, 24,
    41–48, 42, 45, 46, 66, 275
  information garnered by, 42–43, 44,
    47, 256
  in robotics, 60, 61
newcomer rituals, 234
Newman, Tom, 268
New York Times, xii, 56, 108, 133, 145,
    150, 155, 220
Nickell, Jake, 143, 144
99designs, 145, 158, 166, 195
Nivi, Babak, 174
Nokia, 47
Nordstrom, 72
Nye, Bill, 180, 200, 207

"Oatmeal, the" (web comic), 178, 179,
    193, 196, 200
Oculus Rift, 182
O'Dell, Jolie, 238–39
oil-cleanup projects, 247, 250–53, 262,
    263, 264

Olguin, Carlos, 65
1Qbit, 59
operational assets, crowdsourcing of,
    158–60
Orteig Prize, 244, 245, 259, 260, 263
Oxford Martin School, 62

Page, Carl, 135
Page, Gloria, 135
Page, Larry, xiii, 53, 74, 81, 84, 99,
    100, 115, 126, 128, 134–39, 146
  thinking-at-scale strategies of, 136–38
PageRank algorithm, 135
parabolic flights, 110–12, 123
Paramount Pictures, 151
Parliament, British, 245
passion, importance of, 106–7, 113,
    116, 119–20, 122, 125, 134, 174,
    180, 183, 184, 248, 249
  in online communities, 224, 225,
    228, 231, 258
PayPal, 97, 117–18, 167, 201
PC Tools, 150
Pebble Watch campaign, 174, 175–78,
    179, 182, 186, 187, 191, 200, 206,
    208, 209, 210
  pitch video in, 177, 198, 199
peer-to-peer (P2P) lending, 172
Pelton, Joseph, 102
personal computers (PCs), 26, 76
Peter's Laws, 108–14
PHD Comics, 200
philanthropic prizes, 267
photography, 3–6, 10, 15
  demonetization of, 12, 15
  see also digital cameras; Kodak
    Corporation
Pink, Daniel, 79
Pishevar, Shervin, 174
pitch videos, 177, 180, 192, 193, 195,
    198–99, 203, 212
Pivot Power, 19
Pixar, 89, 111
Planetary Resources, Inc., 34, 95, 96, 99,
    109, 172, 175, 179, 180, 186, 189–
    90, 193, 195, 201–3, 221, 228, 230

PRAISE FOR JAMES W. HALL

WITHDRAWN

"A masterful writer."

—James Patterson

"No writer working today . . . more clearly evokes the shadows and loss that hide within the human heart."

—Robert Crais

"The king of the Florida-gothic noir."

—Dennis Lehane

"Delivers taut and muscular stories about a place where evil always lurks beneath the surface."

—Michael Connelly

"I believe no one has written more lyrically of the Gulf Stream since Ernest Hemingway."

—James Lee Burke

"Hall keeps the tension mounting as motives and alliances shift with the foul-scented wind. Even as violence looms, Hall's talent for description adds a balancing, poetical note."

—*Publishers Weekly* on *The Big Finish*

"As ever, Hall is in colorful command of his South Florida setting . . . Compared to other mystery writers, he plays things refreshingly low key, but he's always in control, thriving on the setup as much as the payoff . . . with its nicely observed characters and lively dialogue—and terrific sex scenes—it keeps readers turning the pages."

—*Kirkus Reviews* on *Going Dark*

"A damn good mystery."

—Booklist on Dead Last

"Hall is one of those rare thriller writers who can build character as he ratchets tension, who can do no-holds-barred action scenes with panache and, in the midst of bedlam, never lose sight of nuance. All those skills are on display here, as Hall assembles a full-bodied supporting cast whose stories hold our interest as much as Thorn's attempt to save his son without helping to bring about a South Florida version of Chernobyl. A fine thriller on every level."

—Booklist on Going Dark

"Hall's latest novel, titled Going Dark, proves he's one of the best genre writers working today."

—Alan Cheuse, All Things Considered